Analytic Number Theory
An Introduction

Volumes of the Series published from 1961 to 1973 are not officially numbered. The parenthetical numbers shown are designed to aid librarians and bibliographers to check the completeness of their holdings.

ISBN

0-8053-5801-3	(1)	S. Lang	Algebraic Functions, 1965
0-8053-8703-X	(2)	J. Serre	Lie Algebras and Lie Groups, 1965 (3rd printing, with corrections, 1974)
0-8053-2327-9	(3)	P. J. Cohen	Set Theory and the Continuum Hypothesis, 1966 (4th printing, 1977)
0-8053-5808-0 0-8053-5809-9	(4)	S. Lang	Rapport sur la cohomologie des groupes, 1966
0-8053-8750-1 0-8053-8751-X	(5)	J. Serre	Algèbres de Lie semi-simples complexes, 1966
0-8053-0290-5 0-8053-0291-3	(6)	E. Artin and J. Tate	Class Field Theory, 1967 (2nd printing, 1974)
0-8053-0300-6 0-8053-0301-4	(7)	M. F. Atiyah	K-Theory, 1967
0-8053-2434-8 0-8053-2435-6	(8)	W. Feit	Characters of Finite Groups, 1967
0-8053-3555-4	(9)	Marvin J. Greenberg	Lectures on Algebraic Topology, 1966 (6th printing, with corrections, 1979)
0-8053-3757-1	(10)	Robin Hartshorne	Foundations of Projective Geometry, 1967 (3rd printing, 1978)
0-8053-0660-9	(11)	H. Bass	Algebraic K-Theory, 1968
0-8053-0668-4	(12)	M. Berger and M. Berger	Perspectives in Nonlinearity: An Introduction to Nonlinear Analysis, 1968
0-8053-5208-2 0-8053-5209-0	(13)	I. Kaplansky	Rings of Operators, 1968
0-8053-6690-3 0-8053-6691-1	(14)	I. G. MacDonald	Algebraic Geometry: Introduction to Schemes, 1968
0-8053-6698-9 0-8053-6699-7	(15)	G. W. Mackey	Induced Representation of Groups and Quantum Mechanics, 1968
0-8053-7710-7 0-8053-7711-5	(16)	R. S. Palais	Foundations of Global Nonlinear Analysis, 1968
0-8053-7818-9 0-8053-7819-7	(17)	D. Passman	Permutation Groups, 1968
0-8053-8725-0	(18)	J. Serre	Abelian *l*-Adic Representations and Elliptic Curves, 1968
0-8053-0116-X	(19)	J. F. Adams	Lectures on Lie Groups, 1969
0-8053-0550-5	(20)	J. Barshay	Topics in Ring Theory, 1969
0-8053-1021-5	(21)	A. Borel	Linear Algebraic Groups, 1969
0-8053-1050-9	(22)	R. Bott	Lectures on K(X), 1969
0-8053-1430-X 0-8053-1431-8	(23)	A. Browder	Introduction to Function Algebras, 1969
		G. Choquet	Lectures on Analysis (3rd printing, 1976)
0-8053-6955-4	(24)		Volume I. Integration and Topological Vector Spaces, 1969
0-8053-6957-0	(25)		Volume II. Representation Theory, 1969
0-8053-6959-7	(26)		Volume III. Infinite Dimensional Measures and Problem Solutions, 1969
0-8053-2366-X 0-8053-2367-8	(27)	E. Dyer	Cohomology Theories, 1969

ISBN			
0-8053-2420-8 0-8053-2421-6	(28)	R. Ellis	Lectures on Topological Dynamics, 1969
0-8053-2570-0 0-8053-2571-9	(29)	J. Fogarty	Invariant Theory, 1969
0-8053-3080-1 0-8053-3081-X	(30)	William Fulton	Algebraic Curves: An Introduction to Algebraic Geometry, 1969 (5th printing, with corrections, 1978)
0-8053-3552-8 0-8053-3553-6	(31)	M. J. Greenberg	Lectures on Forms in Many Variables, 1969
0-8053-3940-X 0-8053-3941-8	(32)	R. Hermann	Fourier Analysis on Groups and Partial Wave Analysis, 1969
0-8053-4551-5	(33)	J. F. P. Hudson	Piecewise Linear Topology, 1969
0-8053-5212-0 0-8053-5213-9	(34)	K. M. Kapp and H. Schneider	Completely O-Simple Semigroups: An Abstract Treatment of the Lattice of Congruences, 1969
0-8053-5240-6 0-8053-5241-4	(35)	J. B. Keller and S. Antman, (eds.) O. Loos	Bifurcation Theory and Nonlinear Eigenvalue Problems, 1969 Symmetric Spaces
0-8053-6620-2 0-8053-6621-0	(36)		Volume I. General Theory, 1969
0-8053-6622-9 0-8053-6623-7	(37)		Volume II. Compact Spaces and Classification, 1969
0-8053-7024-2 0-8053-7025-0	(38)	H. Matsumura	Commutative Algebra, 1970 (2nd Edition—cf. Vol. 56)
0-8053-7574-0 0-8053-7575-9	(39)	A. Ogg	Modular Forms and Dirichlet Series, 1969
0-8053-7812-X 0-8053-7813-8	(40)	W. Parry	Entropy and Generators in Ergodic Theory, 1969
0-8053-8350-6 0-8053-8351-4	(41)	W. Rudin	Function Theory in Polydiscs, 1969
0-8053-9100-2 0-8053-9101-0	(42)	S. Sternberg	Celestial Mechanics Part I, 1969
0-8053-9102-9	(43)	S. Sternberg	Celestial Mechanics Part II, 1969
0-8053-9254-8 0-8053-9255-6	(44)	M. E. Sweedler	Hopf Algebras, 1969
0-8053-3946-9 0-8053-3947-7	(45)	R. Hermann	Lectures in Mathematical Physics Volume I, 1970
0-8053-3942-6	(46)	R. Hermann	Lie Algebras and Quantum Mechanics, 1970
0-8053-8364-6 0-8053-8365-4	(47)	D. L. Russell	Optimization Theory, 1970
0-8053-7080-3 0-8053-7081-1	(48)	R. K. Miller	Nonlinear Volterra Integral Equations, 1971
0-8053-1875-5 0-8053-1876-3	(49)	J. L. Challifour	Generalized Functions and Fourier Analysis, 1972
0-8053-3952-3	(50)	R. Hermann	Lectures in Mathematical Physics Volume II, 1972
0-8053-2342-2 0-8053-2343-0	(51)	I. Kra	Automorphic Forms and Kleinian Groups, 1972
0-8053-8380-8 0-8053-8381-6	(52)	G. E. Sacks	Saturated Model Theory, 1972
0-8053-3103-4	(53)	A. M. Garsia	Martingale Inequalities: Seminar Notes on Recent Progress, 1973
0-8053-5664-3 0-8053-5666-5	(54)	T. Y. Lam	The Algebraic Theory of Quadratic Forms, 1973 (2nd printing, with revisions, 1980)
0-8053-6702-0 0-8053-6703-9	55	George W. Mackey	Unitary Group Representations in Physics, Probability, and Number Theory, 1978

MATHEMATICS LECTURE NOTE SERIES *(continued)*

ISBN

0-8053-7026-9	56	Hideyuki Matsumura	Commutative Algebra, Second Edition, 1980
0-8053-0360-X	57	Richard Bellman	Analytic Number Theory: An Introduction, 1980

Other volumes in preparation

Analytic Number Theory
An Introduction

Richard Bellman
University of Southern California
Los Angeles, California

1980

The Benjamin/Cummings Publishing Company, Inc.
Advanced Book Program
Reading, Massachusetts

London · Amsterdam · Don Mills, Ontario · Sydney · Tokyo

Library of Congress Cataloging in Publication Data

Bellman, Richard Ernest, 1920–
  Analytic number theory.

  (Mathematics lecture note series ; 57)
  Includes bibliographies and indexes.
  1. Numbers, Theory of. I. Title.
QA241.B43   512'.7   80-14166
ISBN 0-8053-0360-X

American Mathematical Society (MOS) Subject Classification Scheme (1980): 10-XX

Manufactured in the United States of America

To

Harold N. Shapiro

Friend and fellow adventurer in these domains

Contents

Preface

The purpose of this book is to provide an introduction to analytic number theory. The theme is the mean value of certain important elementary arithmetic functions, the Euler ϕ function, the divisor function, the squarefree function, and the prime divisor function.

Analytic number theory, as the name indicates, is a blend of analysis and number theory. It soon transpires that one needs algebra and geometry.

Let us sketch the contents of the various chapters. In Chapter 1, various estimates used throughout the volume are given. In no other field of mathematics is it as important to have precise estimates. In Chapter 2, various transform methods are given, Fourier series both infinite and finite, the Laplace transform and the Mellin transform. The Mellin transform plays a very important role in number theory, as is indicated. In Chapter 3, some properties of congruences are given. In Chapter 4, some properties of the gamma function are given. In Chapter 5, the Riemann zeta function is introduced. This function plays many roles in number theory. It serves as a generating function for many arithmetical functions of interest and plays an important role in estimating error terms. In Chapter 6, the Poisson summation formula is discussed. This is one of the most powerful methods of obtaining important identities. In Chapter 7, the topic of functional equations is introduced. The utility of this method is shown by application to the theta functions.

In Chapter 8, we turn to various mean values associated with Euler ϕ function. As we show, a representation for this function makes the mean values easy to obtain. In Chapter 9, we consider analogous questions for the divisor function. Here, the corresponding questions are much more difficult.

In Chapter 10, we consider some problems associated with squarefree numbers. What is surprising is how difficult these questions appear to be.

Chapter 11 consists of three unrelated topics. In the first part of this chapter, we consider the prime divisor function, assuming the prime number theorem. In the second part of the chapter, we present the sieve method of Selberg. Finally, in the third part of the chapter, we present some results on the algebraic independence of the elementary arithmetic functions.

Finally, in Chapter 12, we show the utility of Tauberian theory in analytic number theory.

Much has been omitted. We have done this for several reasons. In the first place, there is no point to repeating material which is readily available. In the second place, it would not be in anybody's interest to make the volume too large. In the third place, we have concentrated on the areas in which we have had greatest personal interest. The connections with algebra, particularly, algebraic number theory and group theory, have been hinted at. Many other connections are referred to—either in the text or bibliography and comments.

Any serious student of number theory will want to refer to the following books:

Dickson, L. E., *History of the Theory of Numbers,* Chelsea, New York, 1952.
Landau, E., *Vorlesungen über Zahlentheorie, Teubner, 1921.*
Hardy, G. H., and E. M. Wright, *An Introduction to the Theory of Numbers,* Clarendon Press, Oxford, 5th ed., 1979.

It will be clear how much has been omitted here. In particular, we do not discuss the important subject of trigonometric sums. This topic has been omitted for two reasons. In the first place, the discussion is very technical which makes it unsuitable for an introductory text. In the second place, no brief account of this important area can be given. We have noted a reference to an excellent expository paper by Vinogradov in which many other references will be found.

It will be evident from the text how many open questions there are in analytic number theory. It is thus an ideal field for research.

We presuppose a knowledge of elementary number theory and the rudiments of the theory of functions of a complex variable. Some of the miscellaneous exercises require a bit more. With this background, the book can be used for a one-semester course, or self-study.

Acknowledgments

I wish to thank Tosio Fujiyama, Harold N. Shapiro, and Ernst Straus who read the manuscript and corrected many errata.

Harold N. Shapiro simplified and amplified many proofs and rewrote several sections.

Rebecca Karush typed and retyped the manuscript.

I wish to thank Mohammad Roosta for his help in reading the galley and page proofs and Beth Mooney for preparing the name and subject indexes.

RICHARD BELLMAN

Santa Monica, California

Analytic Number Theory
An Introduction

1. Estimates

1. Introduction

In this chapter, we will obtain some results which we shall use in the rest of the book.

First we shall compare the sum in $\Sigma_{k=0}^{N} f(k)$ with the integral $\int_0^N f(x)dx$ where $f(x)$ is a monotone function. As we shall see, the same technique can be used for lattice points. Then we shall turn to inequalities. The most useful inequality in analysis is the Cauchy-Schwarz inequality.

Following that, we shall prove some simple estimates which we shall use repeatedly. Finally, we shall give a very simple evaluation of the Gauss integral. From the Gauss integral we shall derive the Hecke integral.

2. Comparison of the Sum $\Sigma_{k=0}^{N} f(k)$ with the Integral $\int_0^N f(x)dx$

In this section, we want to obtain some useful elementary estimates. We will use such estimates throughout the book.

Consider the following figure.

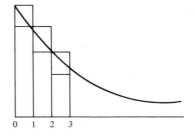

Richard Bellman, Analytic Number Theory: An Introduction   ISBN 0-8053-0452-5

A simple comparison of areas yields the estimates

(1)
$$\sum_{k=1}^{N} f(k) \leqslant \int_{0}^{N} f(x)dx \leqslant \sum_{k=0}^{N-1} f(k).$$

From these inequalities, it is easy to obtain asymptotic results, as will be indicated in the exercises.

Exercises

1. Derive a similar result for the case where $f(k)$ is first increasing and then decreasing.

2. Derive the two-dimensional version of these inequalities.

3. Consider the case where the function of two variables has the form $f(xy)$ and hence derive estimates for the sum $\Sigma_{n=1}^{N} d(n)f(n)$ where $d(n)$ is the number of divisors of n.

4. Consider the case where the function of two variables has the form $f(x^2 + y^2)$ and thus obtain estimates for $\Sigma_{n=1}^{N} r(n)f(n)$ where $r(n)$ is the number of representations of n as the sum of two squares.

3. Lattice Points

The same techniques can be used to count the number of lattice points in a region. This is part of the geometry of numbers which we shall not enter into here.

Exercises

1. Show that the number of lattice points in the region $1 \leqslant xy \leqslant n$ is $\Sigma_{n=1}^{N}$ $[n/x]$ where $[x]$ is the greatest integer less than or equal to x.

2. Show that the number of lattice points in the quadrant of the circle $x^2 + y^2 = n$ is given by $\Sigma_{k=1}^{n} [n^2 - k^2]$.

3. For a convex curve, compare the number of lattice points with the area.

4. Consider the triangular region determined by the lines $x = 0, y = 0, ax + by = 1$. Count the lattice points in this region in two ways and equate.

5. Consider the tetrahedral region determined by the planes $x = 0, y = 0, z = 0, ax + by + cz = 1$. Count the lattice points in this region in three ways and equate.

6. Do the same for a quadrant of an ellipse.

7. What meanings can be ascribed to the statement "The probability that two integers are relatively prime is $6/\pi^2$?." (Lehmer, Bellman-Shapiro).

4. Convergence

We can do much better in many cases, let us write

$$(1) \qquad \sum_{k=0}^{N} f(k) - \int_{0}^{N+1} f(x)dx = \sum_{k=0}^{N} \left(f(k) - \int_{k}^{k+1} f(x)dx \right).$$

Let us now use the elementary result

$$(2) \qquad f(k) - f(x) = \int_{x}^{k} f'(y)dy.$$

We thus have

$$(3) \qquad \sum_{k=1}^{N} \left(f(k) - \int_{k-1}^{k} f(x)dx \right) = \sum_{k=1}^{N} \int_{k-1}^{k} f'(y)dy.$$

In many important cases, the function $f'(y)$ is monotone. Consequently, we see that if $\int_{0}^{\infty} f'(y)dy$ converges, we have the limit.

$$(4) \qquad \lim_{n\to\infty} \left(\sum_{k=0}^{N} f(k) - \int_{0}^{N} f(x)dx \right) = c.$$

Exercises

1. Show that $e^x - 1 \leqslant xe^x$, if $x \geqslant 0$.

2. Show that $1 - e^{-x} \geqslant 0$, if $x \geqslant 0$.

3. Show that $1 + x + \cdots + x^n = (1 - x^{n+1})/(1 - x)$.

4. Hence, show that

$$\frac{1}{1-x} = 1 + x + \cdots + \frac{x^n}{1-x}$$

5. Hence, by integrating, show that

$$\log \frac{1}{1-x} = x + \frac{x^2}{2} + \cdots + \int_0^x \frac{y^n}{1-y}\,dy.$$

6. Using the representation

$$\log(1+x) = \int_0^x \frac{dy}{1+y},$$

show that $(1 + x/n)^n$ is monotonically increasing in n.

7. Assume that $f(h)$ is monotone decreasing and convex. Obtain a better upper bound using the trapezoid formed by connecting k and $k + 1$.

8. Obtain a lower bound by using the trapezoid formed by taking the tangent at $(k, f(k))$.

9. Obtain a lower bound by using the trapezoid formed by taking the tangent at $(k + 1/2, f(k + 1/2))$.

5. Euler's Constant

The most important example is Euler's constant. We have

(1)
$$\lim_{N \to \infty} \left(\sum_{n=1}^N \frac{1}{n} - \log N \right) = \gamma.$$

The values of γ is $0.5771 \cdots$. Most likely, γ is transcendental, but it has not even been shown that γ is irrational. The question of the arithmetic nature of γ is one of the outstanding problems of the theory of numbers.

We shall encounter the Euler constant in Chap. 4.

Exercises

1. Find the asymptotic behavior of $\sum_{n=1}^\infty n^a e^{-nx}$ as $x \to 0$.

2. Obtain a two-dimensional generalization of the formula above.

3. Find the asymptotic behavior of

$$\sum_{n=2}^N \frac{1}{n \log n}.$$

6. Inequalities

In this section, we will derive the Cauchy-Schwarz inequality.
We begin with the fundamental result

(1) $$(a - b)^2 \geqslant 0.$$

Multiplying out, we obtain

(2) $$a^2 + b^2 \geqslant 2ab.$$

Inwards, the arithmetic mean of two positive quantities is greater than the geometric mean.
We now take the fundamental inequality in (2) and set

(3)
$$a = a_n \Big/ \left(\sum_{n=1}^{N} a_n^2 \right)^{1/2},$$

$$b = b_n \Big/ \left(\sum_{n=1}^{N} b_n^2 \right)^{1/2}.$$

Adding, we obtain the famous inequality of Cauchy-Schwarz

(4) $$\left| \sum_{n=1}^{N} a_n b_n \right| \leqslant \left(\sum_{n=1}^{N} a_n^2 \right)^{1/2} \left(\sum_{n=1}^{N} b_n^2 \right)^{1/2}.$$

This is one of the most useful inequalities of analysis.

Exercises

1. Using (2) above obtain the inequality $a_1^4 + a_2^4 + a_3^4 + a_4^4 \geqslant 4a_1 a_2 a_3 a_4$.

2. Continuing in this fashion, obtain the inequality for any power of two.

3. By specialization obtain the famous arithmetic geometric mean inequality:
The arithmetic mean of n nonnegative quantities is always greater than or equal
to the geometric mean.

4. Show that the Cauchy-Schwarz inequality may be written

$$\left(\sum_{n=1}^{N} a_n^2 \right)^{1/2} = \max_{\sum_{n=1}^{N} b_n^2 = 1} \left(\sum_{n=1}^{N} a_n b_n \right).$$

where the max is taken over;

5. From the foregoing representation derive the Minkowski inequality

$$\left(\sum_{n=1}^{N} (a_n + b_n)^2 \right)^{1/2} \leqslant \left(\sum_{n=1}^{N} a_n^2 \right)^{1/2} + \left(\sum_{n=1}^{N} b_n^2 \right)^{1/2}.$$

6. Prove the arithmetic mean-geometric mean inequality for the case $n = 4$ by showing that the difference is nonnegative.

7. Establish the corresponding result for any power of 2.

8. By specialization, obtain a corresponding result for general n.

7. Continuous Version

The continuous version of the Cauchy-Schwarz inequality is also very useful. We begin as before with the inequality

(1) $(f - g)^2 \geqslant 0.$

Multiplying out, and integrating we have

(2) $\int f^2 + \int g^2 \geqslant 2 \int fg.$

This shows that the product fg is integrable whenever f^2 and g^2 are. Using

(3)
$$f = f / \left(\int f^2 \right)^{1/2},$$

$$g = g / \left(\int g^2 \right)^{1/2},$$

we obtain

(4)
$$\left|\int fg\right| \leqslant \left(\int f^2\right)^{1/2}\left(\int g^2\right)^{1/2}.$$

8. A Technique of Shnirelman

From the Cauchy-Schwarz inequality we have

(1)
$$\left(\sum_{n \epsilon S} a_n\right) \leqslant \left(\sum_{n \epsilon S} 1\right)^{1/2}\left(\sum_{n \epsilon S} a_n^2\right)^{1/2}.$$

Solving, we have

(2)
$$\left(\sum_{n \epsilon S} 1\right)^{1/2} \geqslant \frac{\displaystyle\sum_{n \epsilon S} a_n}{\left(\displaystyle\sum_{n \epsilon S} a_n^2\right)^{1/2}}.$$

By a judicious choice of the a_n, we can obtain a lower bound for the number of elements of a set S.

9. Summation by Parts

A most powerful technique for the estimation of integrals is integration by parts. A similar technique exists for sums, summation by parts.

(1)
$$t_n s_n - t_{n-1} s_{n-1} = t_n(s_n - s_{n-1}) + s_{n-1}(t_n - t_{n-1}).$$

Summing over n, we have the formula

(2)
$$s_N t_N - s_0 t_0 = \sum_{n=1}^{N} t_n(s_n - s_{n-1}) + \sum_{n=1}^{N} s_{n-1}(t_n - t_{n-1}).$$

Exercises

1. Show that we can estimate $\displaystyle\sum_{n=1}^{N} n^a n$ if we have the order of magnitude $\Sigma_{n=1}^{N} a_n$.

2. When can we go in the other direction?

3. Obtain the order of magnitude of $\Sigma_{n=1}^{N} d(n)$.

4. Obtain the order of magnitude of $\Sigma_{n=1}^{N} r(n)$.

5. Use integration by parts $\int_0^x f^{(n)}(y)(x - y)^n dy$ and thus obtain a representation for a Taylor series with a remainder.

6. Obtain a remainder estimate for the function e^x.

7. Obtain an estimate for the remainder of $1/(1 - x)^a$.

8. Show that there is a similar formula for Appell polynomials defined by the relation $p_n'(x) = p_{n-1}(x)$.

9. From the representation

$$(x + y)^s - x^s = s \int_x^{x+y} z^{s-1} dz,$$

derive the estimate

$$\frac{(x + y)^s - x^s}{s} \leqslant yx^{\sigma-1}, \quad \text{where } \sigma \text{ is the real part of } s.$$

for

$$0 \leqslant x, y, \sigma.$$

10. Show that integration by parts may be obtained from summation by parts by going to a suitable limit.

10. The Gaussian Integral

One of the most important integrals in analysis is the Gaussian integral. There are many ways of evaluating it, and we will meet it later on in Chap. 4.
We have

$$(1) \qquad \int_0^\infty \int_0^\infty e^{-(x^2+y^2)} dx\, dy = \int_0^{\frac{\pi}{2}} \int_0^\infty e^{-r^2} r\, dr\, d\theta = \pi/4,$$

upon using polar coordinates.
Thus, we have the famous formula

$$(2) \qquad \int_0^\infty e^{-x^2} dx = \frac{\sqrt{\pi}}{2}.$$

Exercises

1. Show that the change of variable $x = x' + ia$ is legitimate by considering the contour

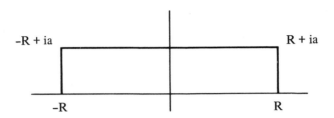

and letting R approach infinity.

2. In this way evaluate the integrals $\int_{-\infty}^{\infty} e^{-x^2} \cos xy\, dx$, $\int_{-\infty}^{\infty} e^{-x^2} \sin xy\, dx$.

11. The Hecke Integral

In the Gaussian integral make the change of variable

(1) $x = y - w/y$.

This has the effect of mapping the interval $(-\infty, \infty)$ into the interval $(0, \infty)$ in a one-to-one fashion.

We see that we have

(2) $dx = dy(1 + w/y^2)$.

Making the change of variable $y = 1/y'$, we see that the second term yields essentially the same integral. We can obtain the same integral by making the change of variable $y = wy'$.

Combining the foregoing results, we have the expression

(3) $\int_0^{\infty} e^{-y-w/y} \dfrac{dy}{y^{1/2}} = e^{-2\sqrt{w}}$.

This integral was used by Hecke to obtain many identities in the analytic theory of numbers.

Exercises

1. Show that the above integral holds if we merely assume that the real part of w is positive.

2. Prove that the function defined by the Hecke integral satisfies a linear second-order differential equation. Hence, evaluate the integral.

Miscellaneous Exercises

1. Establish the identity of Lagrange

$$\sum_i a_i^2 \sum_i b_i^2 - \left(\sum_i a_i b_i\right) = 1/2 \sum_{ij} (a_i b_j - a_j b_i)^2 .$$

2. What is an integral analog of this identity?

3. Show that this identity may be put in determinantal form and thus obtain a higher-dimensional generalization.

4. Show that if $u_{n+m} \leqslant u_n + u_m$, $u_n \geqslant 0$, then $u_{n/m}$ approaches a limit (Fekete).

5. Can the condition $u_n \geqslant 0$ be relaxed?

6. Can the preceding lemma regeneralize to matrices where the relation $A \leqslant B$ means that $B - A$ is nonnegative definite?

Bibliography and Comments

Section 1

A good expository paper is

Davenport, H. *Some Recent Progress in Analytic Number Theory,* J. London Math. Soc. **35** (1960), 135–142.
Rademacher, H., *Trends in Research: The Analytic Number Theory,* Bull. Amer. Math. Soc. **48** (1942), 379–401.
LeVeque, W. J., *Topics in Number Theory, Vol.* 1, Addison-Wesley, Reading, Massachusetts, 1956.

For application of the theory of numbers to other fields, see

Cohen, H., *Some Applied Number Theory,* J. Soc. Indust. Appl. Math. **4.** (1956), 152–167.
Niederreiter, H., *On a Number-Theoretical Integration Method,* Aequationes Math **8** (1972), 304–311.

See also

Lehmer, D. H., *Computer Technology Applied to the Theory of Numbers, Studies in Number Theory,* pp. 117–151, Math. Assoc. Amer. (distributed by Prentice-Hall, Englewood Cliffs, New Jersey, 1969).
Andrews, G. E., *The Use of Computers in Search of Identities of the Rogers-Ramanujan Type, Computers in Number Theory* (Proc. Sci. Res. Council Atlas Sympos. No. 2, Oxford, 1969), pp. 377–387, Academic Press, London, 1971.

Section 3

The study of lattice points is part of the geometry of numbers. Many interesting problems arise in this way, and, as we shall see, there is a connection with the mean values we consider subsequently.

A good account of various methods may be found in

Hardy, G. H. and E. M. Wright, *An Introduction to the Theory of Numbers,* Clarendon Press, Oxford, 1960.

Section 7

For the general theory of inequalities see the books:

Hardy, G. H., J. E. Littlewood and G. Polya, *Inequalities,* Cambridge University Press, London, 1934.
Bellman, R. and E. F. Beckenbach, *Inequalities,* Springer-Verlag, Berlin, 1961, 2nd ed., 1965; 3rd ed., 1970.

Section 8

This is discussed in the book,

Khintchine, A. B., *Three Pearls of Number Theory,* Graylock Press, Albany, New York, 1952.

Section 10

I do not know to whom this technique is due.

2. Transforms

1. Introduction

In this chapter, we want to study some transform methods. These methods are very important in analytic number theory and throughout analysis in general.

In Sec. 2, we study Fourier series. In Sec. 3, we will give the fundamental theorem of Fejer. In Sec. 4, we use this result to establish the result of Parseval. In Sec. 5, we derive a very important Fourier series, which occurs in many parts of analytic number theory.

As a natural generalization of Fourier series we have the Fourier integral. The most important result here is the Parseval-Plancherel theorem.

We turn next to the Laplace transform. In this and the succeeding section we study the inversion formula and the formula for the product. We turn next to the Mellin transform. Again, we are interested in the inversion formula and the formula for the product. Using the Laplace transform, we derive the fundamental identity of Lipshitz, which plays an important role in analytic number theory.

The theory of congruences naturally leads to the study of finite Fourier transforms. In Chap. 12, we shall show that the Laplace transform plays a fundamental role in Tauberian theory. It may also be used to evaluate various integrals and to derive various identities, as we show in the exercises. The Mellin transform plays a fundamental role in the study of Dirichlet series.

2. Fourier Series

The technique of Fourier series is one of the most powerful in analysis. Here, we shall only sketch the outlines of the method.

Richard Bellman, Analytic Number Theory: An Introduction   ISBN 0-8053-0452-5

We begin with the fundamental orthogonality relation,

(1)
$$\int_{-\pi}^{\pi} e^{inx} e^{-imx} dx = 0, \qquad n \neq m,$$
$$= 2\pi, \qquad m = n.$$

Using this relation, we have

(2)
$$f(x) = \sum_{n=-\infty}^{\infty} a_n e^{inx};$$

we have

(3)
$$a_n = \frac{1}{2\pi} \int_{-\pi}^{\pi} f(x) e^{-inx} dx,$$

provided that the series is absolutely convergent.
We can define the Fourier coefficients by means of the relation in (3).
We write

(4)
$$f(x) \sim \sum_{n=-\infty}^{n=\infty} a_n e^{-inx}.$$

The question arises as to when the series on the right actually converges.
Let us write

(5)
$$s_N = \sum_{n=-N}^{n=N} a_n e^{inx}$$
$$= \sum_{n=-N}^{n=N} \left(\frac{1}{2\pi} \int_{-\pi}^{\pi} f(y) e^{-iny} dy \right) e^{inx}$$
$$= \frac{1}{2\pi} \int_{-\pi}^{\pi} f(y) \left(\sum_{n=-N}^{n=N} e^{inx} e^{-iny} \right) dy$$
$$= \frac{1}{2\pi} \int_{-\pi}^{\pi} f(y)(\cdots) dy.$$

The convergence question for Fourier series is delicate because the kernel is not absolutely integrable. Consequently, some condition must be imposed upon $f(y)$. The simplest condition, and the one frequently used in practice, is due to Dirichlet. If $f(y)$ is a bounded variation, then we have convergence. In most applications, that means that $f(y)$ is piecewise differentiable. Then we have convergence at the points of differentiability.

Exercises

1. If $\{\varphi_n(x)\}$ is complete and orthonormal, by which we mean

$$\int_{-\pi}^{\pi} \varphi_n(x)\varphi_m(x)dx = 0, \quad m \neq n,$$

$$= 1, \quad m = n,$$

we have a correspondence

$$f \sim \sum_{n=1}^{\infty} a_n\varphi_n(x)$$

where

$$a_n = \int_{-\pi}^{\pi} f(x)\varphi_n(x)dx.$$

The coefficients are well defined by virtue of the Cauchy-Schwarz inequality if $f \epsilon L^2$.

2. If the sequence is almost-orthogonal, by which we mean

$$\int_{-\pi}^{\pi} \varphi_m(x)\varphi_n(x)dx = a_{mn}$$

where

$$\sum_{m,n=0}^{\infty} a_{mn}x_m x_n \leqslant k \sum_{n=0}^{\infty} x_n^2$$

for some constant k, we also have a correspondence

$$f \sim \sum_{n=1}^{\infty} a_n\varphi_n(x).$$

[See

Bellman, R., *Almost-Orthogonal Series*, Bull. Amer. Math. Soc. **50** (1944), 517–519.
Almost orthogonal series are used in sieve methods.]

3. Show that

$$f e^{iax} \sim \sum_{n=-\infty}^{\infty} a_n e^{inx} e^{-iax}$$

for any integer a.

4. Hence, we have a correspondence

$$f \sim (a_{i-j}).$$

The infinite matrix is called a Toeplitz matrix.

5.
$$\lim_{n \to \infty} \int_{-\pi}^{\pi} \sin nx \left(\frac{1}{\sin x} - \frac{1}{x} \right) dx = 0.$$

6. Hence, evaluate

$$\int_{-\infty}^{\infty} \frac{\sin x}{x} dx.$$

3. Fejer's Theorem

In this section we will establish the famous theorem of Fejer.

This is consistent with the general theme of the book. It may be quite diffi-cult to obtain the behavior at an individual value. However, the behavior of a suitable average, or mean, will be quite smooth.

Let us define

$$(1) \qquad\qquad \sigma_n = \frac{s_0 + s_1 + \cdots + s_N}{N+1}.$$

the arithmetic mean of the partial sums.

It follows that

$$(2) \qquad\qquad \sigma_n = \int_{-\pi}^{\pi} f(y)(\cdots) dy.$$

The important thing is that the kernel is now nonnegative. This nonnegativity makes the discussion of σ_n very easy. We thus have the famous result of Fejer: If $f(y)$ is continuous, σ_n converges uniformly throughout the entire interval.

Exercises

1. Show that

$$\sigma_N = \sum_{k=-N}^{k=N} \frac{1 - |k|}{N+1} a_k e^{ikx}.$$

2. Show that every continuous function over an interval may be uniformly approximated by a trigonometric polynomial.

3. Show that every continuous function over an interval may be uniformly approximated by a polynomial.

4. Show that every continuous function over an interval may be uniformly approximated by a function with an absolutely convergent Fourier series.

5. Show that if f is continuous the condition

$$\int_{-\pi}^{\pi} f(x)e^{inx}\,dx = 0, \quad n = 0, \quad \pm 1, \cdots,$$

implies that $f(x)$ is identically 0.

6. Show that the condition

$$\int_{-\pi}^{\pi} f(x)x^n\,dx = 0, \quad n = 0, 1, \cdots,$$

implies that $f(x)$ is identically 0 if f is integrable.

7. Use Fejer's kernel to establish the value of

$$\int \frac{\sin^2 x}{x^2}\,dx.$$

8. If the series $\Sigma_{n=1}^{\infty} a_n$ has its arithmetic mean of partial sums convergent, and if $\Sigma_{k=1}^{n} ka_k = o(n)$, then the original series is convergent. (This is the original result of Tauber which stimulated Hardy and Littlewood in their research.)

9. Show that if a periodic function is analytic, it has an absolutely convergent Fourier series, and all its derivatives may be calculated from the series.

4. Parseval's Theorem

We have

(1)
$$\int_{-\pi}^{\pi} s_n^2\,dx = \sum_{k=-n}^{n} a_k^2.$$

We wish to extend this result. For this purpose we shall use the foregoing result of Fejer. However, first, we have Bessel's inequality.

(2)
$$\sum_{n=-\infty}^{\infty} |a_n|^2 \leq \frac{1}{2\pi} \int_{-\pi}^{\pi} |f(x)|^2 dx.$$

This asserts that the left side converges whenever the right side exists. To establish this we begin with

(3)
$$\sum_{n=-N}^{N} a_n^2 = \int_{-\pi}^{\pi} f(x) s_n(x) dx$$

and use the Cauchy-Schwarz inequality.

We have

(4)
$$\sum_{k=-N}^{k=N} \left(1 - \frac{|k|}{N+1}\right)^2 a_k^2 = \int_{-\pi}^{\pi} \sigma_N^2(x)\, dx.$$

We know that as $N \to \infty$, on the left, because of the convergence of the series we see that the limit is $\sum_{n=-\infty}^{\infty} a_n^2$. On the right, because of the uniform convergence we have $\int_{-\pi}^{\pi} f^2(x) dx$.

We thus have the formula of Parseval

(5)
$$\sum_{n=-\infty}^{\infty} a_n^2 = \frac{1}{2\pi} \int_{-\pi}^{\pi} f^2(x) dx.$$

valid for any continuous function $f(x)$.

Exercise

1. Obtain Bessel's inequality by determining the minimum of

$$\int_0^T (f^2 - c_1 \phi_1 - \cdots - c_n \phi_n)^2 dt.$$

5. Fourier Series for the Fractional Part

Consider the function $x - [x]$. This represents the fractional part of x, and is an important number-theoretic function. We see that it is periodic of period one. Hence, we have the Fourier series,

$$(1) \qquad x - [x] = \frac{1}{2} + \sum_{n=1}^{\infty} \sin \frac{2\pi n x}{\pi n},$$

as an easy calculation shows.

In the following, this series will be used for several purposes.

Exercises

1. Use Parseval's theorem to evaluate

$$\sum_{n=1}^{\infty} \frac{1}{n^2}.$$

2. In the same fashion evaluate $\sum_{n=1}^{\infty} 1/n^{2k}$.

6. Finite Fourier Series

In many parts of number theory, we meet a sequence which is periodic. It is convenient to treat such sequences by a finite Fourier series. We write

$$(1) \qquad f(n) = \sum_{k=-N}^{N} a_k e^{2\pi i n k/N}.$$

There is now no question of convergence, nor of the existence of Parseval's relation. We shall not consider finite Fourier series because they are intimately connected with trigonometric sums, which we have agreed not to treat.

Exercises

1. If $f(x)$ has an absolutely convergence Fourier series we have

$$f(k/N) = \sum_{r=-\infty}^{\infty} a_r e^{2\pi i r k/N}$$

$$= \sum_{r=-N}^{\infty} e^{2\Pi i r k/N} (a_r + a_{r+N} + \cdots).$$

2. Find the finite Fourier series for $f(k/N)e^{2\Pi ika/N}$.

7. The Fourier Transform

Let us now consider the case when the interval is infinite. If we proceed formally, we see that the series becomes an integral. This procedure may be made rigorous. However, it is easier to begin with the expression for the infinite integral

(1) $$F(y) = \int_{-\infty}^{\infty} f(x)e^{ixy}dx.$$

This expression is called the Fourier integral. We are interested here in two features, the inversion formula and the Parseval relation. In order to make this symmetrical, the integral is often written

(2) $$F(y) = \frac{1}{\sqrt{2\pi}} \int_{-\infty}^{\infty} f(x)e^{ixy}dx.$$

Then we have the elegant inversion formula

(3) $$f(x) = \frac{1}{\sqrt{2\pi}} \int_{-\infty}^{\infty} F(y)e^{-ixy}dy.$$

8. Parseval-Plancherel Relation

We have

(1) $$\int_{-\infty}^{\infty} |F(y)|^2 dy = \int_{-\infty}^{\infty} |f(x)|^2 dx.$$

We will employ this result in Chap. 12. A proof of this fundamental result may be obtained in various ways as the book by Titchmarsh shows.

9. The Laplace Transform

We write

(1) $$L(f) = F(s) = \int_{0}^{\infty} e^{-st}f(t)dt, \ t > 0.$$

We may regard this as a continuous version of a generating function. We have the inversion formula

$$(2) \qquad f(t) = \frac{1}{2\pi i} \int_C e^{st} F(s) ds.$$

Here C is a contour which lies to the right of the singularities of F. We shall take C to be a straight line. This inversion formula allows us to use the theory of a complex variable.

We may regard the Laplace transform as a special case of the Fourier transform, or conversely.

Exercises

1. Use the Parseval-Plancherel formula to establish the result

$$\int_0^\infty f(t) e^{-st} dt = \int_0^\infty \frac{F(x)}{s + ix} dx,$$

where now F is the Fourier transform of f.

2. Use the result $(1 + ix/n)^n \to e^{ix}$ to obtain the post-Widder inversion formula. (Other proofs of this inversion formula will be found in the books by Doetsch and Widder cited in the bibliography.)

3. Consider the maximum transform defined by the relation $\max_{x \geqslant 0} f(x) e^{-xy} = F(y)$. Show under appropriate conditions that we have $f(x) = \min_{y \geqslant 0} F(y) e^{xy}$.

See,

Bellman, R. and W. Karush, *On a New Functional Transform in Analysis: The Maximum Transform,* Bull. Amer. Math. Soc. **67** (1961), 501–503.

——, *Mathematical Programming and the Maximum Transform,* J. Soc. Indust. Appl. Math. **10** (1962), 550–567.

——, *On the Maximum Transform and Semigroups of Transformations,* Bull. Amer. Math. Soc. **68** (1962), 516–518.

——, *On the Maximum Transform,* J. Math. Anal. Appl. **6** (1963), 67–74.

——, *Functional Equations in the Theory of Dynamic Programming-*XII: *An Application of the Maximum Transform,* J. Math. Anal. Appl. **6** (1963), pp. 155–157.

4. Consider the function defined by the relation
$$h(x) = \max_{x_1 + x_2 = x} (f(x_1) + g(x_2)).$$

Show that

$$\max_{x \geqslant 0} h(x) = \max_{x \geqslant 0} f(x) + \max_{x \geqslant 0} g(x).$$

(This relation plays an important role in some problems of dynamic programming. See
Bellman, R. and S. Dreyfus, *Applied Dynamic Programming*, Princeton University Press, Princeton, NJ, 1962.

10. The Product Formula

Consider the function

(1)
$$h(x) = \int_0^x f(x - y)g(y)dy.$$

Then, we have the fundamental result of Borel

(2)
$$L(h) = L(f)L(g).$$

The function $h(x)$ is usually called the convolution of f and g. The validity of Borel's formula may be established under various conditions on f and g.

11. The Identity of Lipshitz

Let us now derive a fundamental identity. This identity plays an important role in the theory of numbers, as we will show. We begin with the Fourier expansion of $x - [x]$ derived above.

We have

(1)
$$\int_0^\infty e^{-sx}(x - [x])dx = \int_0^\infty e^{-sx}\left(\frac{1}{2} - \sum_{n=1}^\infty \frac{\operatorname{Sin} 2\pi nx}{n\pi}\right)dx.$$

On the left, we write

(2)
$$\sum_{n=0}^\infty \int_n^{n+1} e^{-sx}(x - n)dx;$$

performing this integration, and interchanging the summation and integration on the right, the legitimacy we will leave to the reader, we obtain the identity

(3)
$$\frac{1}{1 - e^s} = \frac{1}{s} + \frac{1}{2}\sum_{n=1}^\infty \frac{2s}{s^2 + 4n^2\pi^2}, s \neq 0.$$

Exercises

1. Obtain the partial fraction expansion of $1/(1 - x^n)$. (An interesting generalization of partial fractions was obtained by Jacobi. See Jacobi, C. G., Gesamnaelte Werke.

2. Using this result, with a suitable choice of x and letting n approach infinity, derive the foregoing identity.

12. The Mellin Transform

Let us now introduce the Mellin transform. We write

$$(1) \qquad\qquad M(f) = F(s) = \int_0^\infty t^{s-1} f(t)\, dt.$$

We have the inversion formula

$$(2) \qquad\qquad f(t) = \frac{1}{2\pi i} \int_C t^{-s} F(s)\, ds,$$

where again C is a contour to the right of the singularities $F(s)$. We shall employ this formula extensively.

13. The Product Formula

Consider the function

$$(1) \qquad\qquad h(x) = \int_0^\infty \frac{f(t)g(x/t)\, dt}{t}.$$

Then we have formally,

$$(2) \qquad\qquad M(h) = M(f)M(g).$$

This result may be established under various conditions on f and g. It enables us to evaluate many integrals as we shall see.

Miscellaneous Exercises

1. Show that if we consider a double Fourier series and if the function is of the form $f(x_1^2 + x_2^2)$, then the Fourier transform is of the same form.

2. Obtain in this way the Bessel transform.

3. Use the Laplace transform on the integral $\int_{-\infty}^{\infty} e^{-tx^2} dx$ and thus evaluate the Gaussian integral.

4. Show that the Fourier expansion may be obtained from integral equation theory by considering the equation

$$\lambda\phi(x) = \int_0^{2\pi} f(x - y)\phi(y)\, dy.$$

Bibliography and Comments

Section 2

In many cases the exponents are incommensurable. This leads to the theory of almost-periodic functions. See the book

Bohr, H., *Almost Periodic Functions,* Chelsea, New York, 1947.

For applications to some questions of astronomy, see the papers

Weyl, H., Amer. J. Math.

A detailed study of Fourier series will be found in the book

Zygmund, A., *Trigonometric Series,* Cambridge University Press, London, 1968.

A detailed study of orthogonal series will be found in the book

Kacmarz, R., and H. Steinhaus, *Theorie der Orthogonal reihen,* Warszawa-Lwow, Z subwencji, Funduszu Kultury narodowej, 1935.

The origin of the theory of Fourier series is in the theory of heat conduction. The original method of Fourier was quite complicated.

For a history of Fourier series, see the monograph by R. E. Langer.

Section 3

For the theory of summability, see the book

Hardy, G. H., *Theory of Summability,* The University Press, Cambridge, England 1945.

Section 5

The theory of the distribution of the fractional part was started by Kronecker. See the book

Koksma, J. F., *Gleichwerteching,* Ergebnisse der Mathematik, Springer-Verlag, 1931.

The bracket function is one of the most interesting functions of number theory. It can be used for many purposes and many interesting identities exist for it.

We have made no attempt to collect all the reciprocity results which occur in number theory. As we have shown in the text, many of these can be obtained from modular transformations.

Iseki, Shô, *The Transformation Formula for the Dedekind Modular Function and Related Functional Equations,* Duke Math. J. 24 (1957), 653–662.

Mikolas, M., *On Certain Sums Generating the Dedekind Sums and Their Reciprocity Laws,* Pacific J. Math. **7** (1957), 1167–1178.

Rademacher, H., *Some Remarks on Certain Generalized Dedekind Sums,* Acta Arith. **9** (1964), 97–105.

Carlitz, L., *Generalized Dedekind Sums,* Math. Z. **85** (1964), 83–90.

——, *A Theorem on Generalized Dedekind Sums, Acta Math.* **11** (1965), 253–260.

——, *Linear Relations Among Generalized Dedekind Sums,* J. Reine Angew. Math. **220** (1965), 154–162.

Vasu, A. C., *A Generalization of Brauer-Rademacher Identity,* Math. Student **33** (1965), 97–101.

Shipp, R. D., *Table of Dedekind Sums,* J. Res. Nat. Bur. Standards Sect. B. **69B** (1965), 259–263.

Carlitz, L., *A Three-Term Relation for the Dedekind-Rademacher Sums,* Publ. Math. Debrecen **14** (1967), 119–124.

Shader, L. E., *The Unitary Brauer-Rademacher Identity,* (Italian summary), Atti Accad. Naz. Lincei Rend. Cl. Sci. Fis. Mat. Natur., **48** (8) (1970), 403–404.

Subbarao, M. V., *The Brauer-Rademacher Identity,* Amer. Math. Monthly **72** (1965), 135–138.

Apostol, T. M., *Generalized Dedekind Sums and Transformation Formulae of Certain Lambert Series,* Duke Math. J. **17** (1950), 147–157.

Mordell, L. J., *The Reciprocity Formula for Dedekind Sums,* Amer. J. Math. **73** (1951), 593–598.

Carlitz, L., *Some Sums Analogous to Dedekind Sums,* Duke Math. J., **20** (1953), 161–171.

——, *Some Theorems on Generalized Dedekind Sums,* Pacific J. Math, **3** (1953), 513–522.

——, *The Reciprocity Theorem for Dedekind Sums,* Pacific J. Math., **3** (1953), 523–527.

——, *Dedekind Sums and Lambert Series,* Proc. Amer. Math. Soc. **5** (1954), 580–584.

——, *A Note on Generalized Dedekind Sums,* Duke Math. J. **21** (1954), 399–403.

——, *A Further Note on Dedekind Sums,* Duke Math. J. **23** (1956), 219–223.

Rademacher, H., *Generalizations of the Reciprocity Formula for Dedekind Sums,* Duke Math. J. **21** (1954), 391–397.

Grosswald, E., *Dedekind-Rademacher Sums and Their Reciprocity Formula,* J. Reine Angew. Math. **251** (1971), 161–173.

——, *Dedekind-Rademacher Sums,* Amer. Math. Monthly **78** (1971), 639–644.

Goldstein, L. J., *Dedekind Sums for a Fuchsian Group.* I, Nagoya Math. J. **50** (1973), 21–47.

Rademacher, H. and A. Whiteman, *Theorems on Dedekind Sums,* Amer. J. Math. **63** (1941), 377–407.

Grimson, R. C., *Reciprocity Theorem for Dedekind Sums,* Amer. Math. Monthly **81** (1974), 747–749.

Section 6

Finite Fourier series are very important in analytic number theory. For an application, see

Whiteman, A. L., *Finite Fourier Series and Equations in Finite Fields,* Trans. Amer. Math. Soc. **74** (1953), 78–98.

Section 7

The theory of the Fourier transform is a little more involved because of the infinite interval. See the book

Titchmarsh, E. C., *The Fourier Integral,* Clarendon Press, Oxford, 1959.

In this book, numerous results concerning convergence are given, as well as results on the inversion formula and Parseval's theorem.

Section 9

The Laplace transform is discussed in detail in the three books

Bellman, R. and K. L. Cooke, *Differential-Difference Equations,* Academic Press, New York, 1963.
Doetsch, G., *Theorie und Anwendung der LaPlace-Transformation,* Dover, 1943.
Widder, D. V., *The Laplace Transform,* Princeton University Press, Princeton, New Jersey, 1941.

The problem of numerical inversion is very important in many scientific applications. This problem is considerably complicated by the fact that the Laplace inverse is an unbounded operator. Consequently, we must add some condition on the function $f(s)$. See the book

Bellman, R., R. Kalaba and J. Lockett, *Numerical Inversion of the Laplace Transform,* American Elsevier, New York, 1966.

See

Gesztelyi, E., *The Application of the Operational Calculus in the Theory of Numbers, Number Theory* (Colloq., Janos Bolyai Math. Soc., Debrecen, 1968), pp. 51–104, North-Holland, Amsterdam, 1970.

Section 11

This is one of the fundamental identities of analytic number theory. It has been generalized by Hecke and Siegel.

This identity was first pointed out by Lipshitz. As we shall indicate in Chap. 6, there are important generalizations.

This result can be obtained from general results in the theory of the complex variable, the expansion of Mittag-Leffler.

Section 12

The Mellin transform can be discussed as a special case of either the Laplace transform or the Fourier integral. See the book above by Titchmarsh for a discussion of rigorous aspects of this integral.

3. Congruences

1. Introduction

In this Chapter we want to introduce the residue notation following Gauss. We shall recall various facts which we shall use in the following. This is not intended to be an introduction to congruences, rather a review.

In the second section, we introduce the congruence notation. What makes this notation so powerful is the fact that the congruences can be manipulated like ordinary equations. In Sec. 3, we prove the fundamental theorem of Fermat. With the aid of this relation in Sec. 4 we can treat the linear equation, and give a simple representation for the solution. In Sec. 5, we discuss some facts about the quadratic equations. Here we say a word or two about the famous quadratic reciprocity law. In Sec. 6, we turn to the general polynomial congruence. Here algebraic number theory is required, and we shall just state the results we need.

Finally, we point out the connection between congruences and trigonometric sums.

2. The Congruence Notation

We write

$$(1) \qquad\qquad a \equiv b(p),$$

read "a is congruent to b modulo p".

This means that $a - b$ is divisible by p. Congruences may be manipulated like ordinary equations.

Richard Bellman, Analytic Number Theory: An Introduction   ISBN 0-8053-0452-5

Exercises

1. Show that a necessary and sufficient condition that the congruence

$$\sum_j a_{ij}x_j \equiv 0(p)$$

is that the determinant $|a_{ij}| \equiv 0(p)$.

2. If this condition is satisfied, what is the solution?

3. Show that the polynomial determined by the conditions

$$\sum_{k=0}^{p-1} k^r p(k) \equiv 0(p), \quad 0 \leqslant r < p - 1,$$

is uniquely determined.

4. Show that these polynomials are orthogonal.

5. Show that these polynomials satisfy a 3-term recurrence relation and find this recurrence relation. (These polynomials are analogs of the Legendre polynomials.)

6. Consider the nonzero residues modulo p. Show that by means of the Euclidean algorithm we can determine a reciprocal of each nonzero residue.

7. Show that the nonzero residues form a group, if p is a prime.

3. Fermat's Theorem

In this section we want to establish the fundamental result of Fermat. Naturally, there are many proofs of this result, and we shall indicate another one in the exercises.

Let us begin with a relation

(1) $a_i a \equiv b_j(p).$

Here a_i, a, and b_j are residues modulo p. We take initially p to be a prime. Here a_i, a, and b_j are nonzero residues.

If the a_i are different the b_j must be different. This follows from the fact that there can be no zero divisors modulo p. We now take the relation in (1) and multiply over all the residues. The product on the left must be the same as the product on the right. The result is

(2) $a^{p-1} \equiv 1(p),$

the famous theorem of Fermat.

We can go a bit further. The relation in (2) is equivalent to saying that

(3)
$$a^{p-1} = 1 + kp.$$

Raising both sides to the pth power, we have

(4)
$$a^{p(p-1)} \equiv 1\,(p^2).$$

Proceeding inductably, we have

(5)
$$a^{p^k(p-1)} \equiv 1(p^{k+1}).$$

It follows from the unique factorization theorem that we have the extended form of the theorem of Fermat

(6)
$$a^{\phi(n)} \equiv 1(n).$$

Here, $\phi(n)$ is the Euler ϕ function studied in Chap. 8.

Exercises

1. Consider the set of powers $1, a, a^2, \cdots$. Show that two powers must be the same and thus we have the congruence

$$a^k \equiv 1(p), \qquad \text{where } k \text{ may depend on } a.$$

2. Show that if $a^d \equiv 1(p)$, then $d|(p-1)$.

3. Show that the nonzero residue modulo p form a group.

4. Derive Exercise 1 from this fact.

5. Consider the two by two matrices whose elements are the residues modulo p. Show that the nonsingular matrices form a group.

6. Show that the matrices whose determinant is one form a group.

7. What are the orders of these two groups?

8. Obtain a Fermat theorem for matrices.

9. Consider the recurrence relation

$$u_{n+1} \equiv a_{11}u_n + a_{12}v_n(p),$$
$$v_{n+1} \equiv a_{21}u_n + a_{22}v_n(p).$$

Show that we have a Fermat theorem.

10. Consider the recurrence relation

$$u_{n+2} \equiv au_{n+1} + bu_n(p) .$$

Show that the sequence (u_n) is periodic and satisfies a Fermat relation.

11. Prove the result of Lagrange that the order of a subgroup must be a divisor of the order of the group.

12. Consider the set $1, a, a^2, \cdots, a^{p-2}$. Since there are $p - 1$ numbers, there is a possibility that all the residues are given this way for a suitable a. If such an a exists, it is called a primitive root. How many primitive roots are there?

13. Show that if

$$f(z) = (az + b)/(cz + d),$$

then the iteration of $f(z)$ has the same form.

14. Consider the case where a, b, c, and z are residues modulo p. Show that the iteration is equivalent to matrix multiplication.

15. Hence, derive a Fermat's theorem for the iterates of $f(z)$.

4. The Linear Equation

Consider the equation

(1) $an + b \equiv 0(p).$

By a simple change of variable, we may take the equation in the form

(2) $an \equiv b(p).$

Since there are no zero divisors, there is at most one solution. Using Fermat's theorem, we may exhibit a solution in the form

(3) $n \equiv a^{p-2} b(p).$

Here, as usual, p denotes a prime.

Exercises

1. What is a number of solutions of $an \equiv b(p)$ for $1 \leqslant n \leqslant x$?

2. Consider the equation $AX \equiv B(p)$, where A, X, and B are matrices.

5. The Quadratic Equation

Consider the equation

$$(1) \qquad an^2 + bn + c \equiv 0(p).$$

By the usual change of variable, we may take the equation in the simpler form

$$(2) \qquad n^2 \equiv d(p).$$

The first fact we observe is that this equation is not always soluble. Thus, for example, the equation

$$(3) \qquad n^2 \equiv 2(5),$$

is not soluble. Thus, we have the interesting question of what equations are soluble. If the equation in (2) is soluble, we call d a quadratic residue. If the equation is not soluble we call d a quadratic nonresidue.

It turns out that there is a relation between the solubility of (2) and

$$(4) \qquad n^2 \equiv p(d).$$

This is the substance of the famous law of quadratic reciprocity, conjectured by Euler on the basis of examining a large number of cases, and proved by Gauss. In Chap. 6, we give an extensive generalization of this law involving Gauss sums.

Exercises

1. Prove that the quadratic residues form a group.

2. If the congruence $n^2 \equiv a(p)$ has a solution then $(p - 1)/a^2 \equiv 1(p)$.

3. Is the converse true?

4. Is the product of two quadratic nonresidues always a non-residue?

6. The General Polynomial Equation

Let us now turn to the general polynomial equation

$$(1) \qquad f(n) \equiv 0(p).$$

This equation needs algebraic number theory for its study. Let us cite the following results which we will use. Let $\rho(p)$ denote the number of solutions of this congruence. Then we have

$$(2) \qquad \sum_{n=1}^{N} \frac{\rho(n)}{n} \sim \log N$$

as N becomes infinite. We also have

$$(3) \qquad \sum_{p \leqslant N} \frac{\rho(p)}{p} \sim \log \log N.$$

Here we assume that $f(x)$ is irreducible.

Here, the summation is over the primes less than or equal to N. For the quadratic case, these relations may be established using the Legendre symbol.

Exercises

1. Show that every polynomial may be considered the characteristic polynomial of a matrix.

2. Show that $\rho(nm) = \rho(n)\rho(m)$, if $(n, m) = 1$.

3. Show that we have the factorization

$$n^p - 1 \equiv (n-1)(n-2) \cdots (n-p+1)(p).$$

7. Connection with Trigonometric Sums

We begin with the relation

$$\sum_{k=0}^{p-1} e \frac{2\pi i k r}{p} = 0 \quad \text{if } p \times r,$$

$$= p \quad \text{if } p \,|\, r.$$

We thus have a method for converting a congruence into the evaluation of a trigonometric sum. Thus, for example, we have the expression

$$(2) \qquad \sum_{n=0}^{p-1} \sum_{k=0}^{p-1} e \frac{2\pi i k f(n)}{p} = P\rho(n).$$

Exercises

1. Show that the number of solutions of $x^2 + y^2 \equiv 1(p)$ may be expressed by trigonometric sums.

2. Write

$$\left| \sum_{k=0}^{p-1} e^{\frac{2\pi i k^2 r}{p}} \right|^2 = \sum \sum e^{\frac{2\pi i r(k^2 - l^2)}{p}}$$

and thus show that

$$\left| \sum_{k=0}^{p-1} e^{\frac{2\pi i k^2 r}{p}} \right|^2 = p. \quad \text{(Gauss)}$$

3. Obtain a generalization of the result in the text for simultaneous congruences.

4. Obtain a representation of the number of solutions of the relation

$$\sum_j a_{ij} n_j \equiv 0(p).$$

5. Show that the congruence $f(n) \equiv 0(p)$ is equivalent to a congruence of the above form, and thus obtain a representation for $\rho(n)$. (This method was sketched in,

R. Bellman, *A Note on the Solution of Polynomial Congruences*, Boll. Un. Mat. Ital. **19** (1964), 60–63.)

6. Show that

$$\left| \sum_{n=1}^{k} e^{\frac{2\pi i n^2 r}{p}} \right| = 0(\sqrt{p} \log p). \quad \text{(Polya)}$$

Miscellaneous Exercises

1. Show that two polynomials of the form $x^2 + y^2$ have a product of the same form.

2. Show that this relation is equivalent to the addition formulas for the cosine and sine.

3. Show that we have the factorization $x^3 + y^3 + z^3 - 3xyz = (x + y + x)$ $(x^2 + y^2 + z^2 - xy - xz - yz)$.

Even more dramatic, as H. N. Shapiro points out, $2(x^3 + y^3 + z^3 - 3xyz) = (x + y + z) ((x - y)^2 + (y - z)^2 + (z - x)^2)$, which shows that the real zeros of $x^3 + y^3 + z^3 - 3xyz$ are just the plane $x + y + z = 0$ and the line $x = y = z$.

4. Show that this relation can be used to prove that the cubic congruence $an^3 + bn^2 + cn + d \equiv 0(p)$ is equivalent to a simple cubic and quadratic congruence.

5. Show that the congruence $an^4 + bn^3 + cn^2 + dn + e \equiv 0(p)$ can be solved in terms of a simple fourth degree, third degree and quadratic congruence.

6. Show that a similar result does not hold for the congruence of the fifth degree. (For this result, a knowledge of Galois theory is required.)

7. Show that the result of Exercise 1 may be obtained by observing that

$$x^2 + y^2 = (x + iy)(x - iy).$$

8. Show that the result of Exercise 1 may be obtained by observing that

$$x^2 + y^2 = \begin{vmatrix} xy \\ -yx \end{vmatrix}.$$

9. Show that we have the correspondence

$$x + iy \sim \begin{vmatrix} xy \\ -yx \end{vmatrix}.$$

10. Show that the product of two expressions of the form

$$x^3 + y^3 + z^3 + 3xyz$$

is again an expression of the same form.

11. Show that the product of two expressions of the form $x^2 - 2y^2$ is again of the same form.

12. Given two polynomials of the form $x^2 + y^2$, there is always a third such that the first is the product of the second and third.

13. Derive Exercise 5 from the addition formula for the cosine and sine.

14. Derive Exercise 5 from the representation $x^2 + y^2 = (x + iy)(x - iy)$.

15. Use the representation

$$x_1^2 + x_2^2 + x_3^2 + x_4^2 = (x_1 + ix_2 + jx_3 + kx_4)(x_1 - ix_2 - jx_3 - kx_4),$$

where i, j, and k are quarternions to prove that the product of two polynomials which are the sum of four squares is always the sum of four squares.

16. Derive Exercise 5 from the fact that $x^2 + y^2$ is the determinant of the matrix

$$\begin{pmatrix} xy \\ -yx \end{pmatrix}$$

17. Consider the congruence $x^2 + y^2 \equiv 1(p)$. Associate with a solution of this congruence the complex number $x + iy$. Prove that the product of two such numbers yields another solution.

18. Prove that the quotient of two such numbers yields another solution.

19. Prove that these numbers form a group.

20. What is the order of this group?

21. Establish similar results for the congruence

$$x_1^2 + x_2^2 + x_3^2 + x_4^2 \equiv 1(p).$$

22. Consider the congruence $x_1^2 + x_2^2 + x_3^2 \equiv 1(p)$. Show that we can obtain solutions by adding the conditions $x_4 \equiv 0(p)$.

23. Show that no similar result exists for a sum of three squares. (A famous result of Hurwitz asserts that we only have a product formula for two, four, or eight squares.

24. Consider the congruence $y^2 \equiv ax^3 + bx^2 + cx + d(p)$. Show that the addition formula for Weierstrassian elliptic functions can be used to generate solutions.

25. Consider the congruence $y^2 \equiv ax^4 + bx^3 + cx^2 + dx + e(p)$. Show that the addition formula for Jacobian elliptic functions can be used to generate solutions. (The fact that elliptic functions enter into the consideration of these congruences is not surprising. Hasse showed that the Riemann hypothesis in fields of characteristic p can be treated by elliptic functions in the cubic case. Weil considered the general case using Abelian functions.

26. Show how to generate solutions of the congruence $x^2 - 2y^2 \equiv 1(p)$.

27. Show how to generate solutions of the congruence $x^3 + y^3 + z^3 - 3xyz \equiv 1(p)$.

28. Show that $x^n - 1$ and $x^m - 1$ have only the factor $x - 1$ in common if n and m are relatively prime.

29. Prove that the numbers c for which the congruence $x^2 + y^2 \equiv c(p)$ has a solution form a group.

30. Prove that there exist values of p for which the congruence $x^2 + y^2 \equiv 0(p)$ has a nontrivial solution.

31. Consider that the quantities of the form $a + ib$, where a and b are not both simultaneously zero, form a group for some value of p and not for others. For what values of p do they form a group?

32. Does there exist a polynomial $g(n)$ such that we have the congruence $g(an + bn^2) \equiv ag(n)\ (p)$?

33. Do we have a Fermat's theorem for the iterates of $an + bn^2$?

34. Does a primitive root exist for quantities of the form $a + ib$?

35. Consider the 2×2 matrices where the elements are residues modulo p. Show that multiplication is not necessarily commutative.

36. Show that the 2×2 matrices cannot be generated as powers of a fixed matrix. In other words, there is no analog for matrices of a primative root.

38. Consider the general polynomial congruence. Multiply by the quantities x^k, $0 \leqslant k \leqslant p - 2$. Regard the resulting system as a set of equations for the quantites x^k. Obtain in this way a necessary determinantal criterion for a solution.

39. Is this condition sufficient?

Bibliography and Comments

Section 1

For an extensive treatment of congruences, see

G. H. Hardy and E. M. Wright, *An Introduction to the Theory of Numbers,* Clarendon Press, Oxford, 1960.

An extension generalization of the notion of residue is the Galois, or finite field. For older results, see the book

L. E. Dickson, *Linear Groups with an Exposition of the Galois Field Theory,* Dover, New York, 1958.

For more recent results, see the paper by L. Carlitz and his students. See also,

K. E. Kloss, *Some Number-Theoretic Calculations,* J. Res. Nat. Bur. Standards Sec. B. **69B** (1965), 335–336.

C. E. Froberg, *On Some Number-Theoretical Problems Treated with Computers, Computers in Mathematical Research,* pp. 84–88, North-Holland, Amsterdam, 1968.

Section 2

See,

H. N. Shapiro, *Note on a Problem in Number Theory,* Bull. Amer. Math. Soc. **54** (1948), 890–893.

There are many interesting questions concerning sequences, see

M. Ward, *Memoir on Elliptic Divisibility Sequences,* Amer. J. Math. **70** (1948), 31–74.

Brenner, J. L., *Linear Recurrence Relations,* Amer. Math. Monthly **61** (1954), 171–173.

Durst, L. K., *Exceptional Real Lehmer Sequences,* Pacific J. Math. **9** (1959), 437–441.

Chowla, S., M. Dunton and D. J. Lewis, *Linear Recurrences of Order Two,* Pacific J. Math. **11** (1961), 833–845.

Howard, F. T., *A Property of a Class of Nonlinear Difference Equations,* Proc. Amer. Math. Soc. **38** (1973), 15–21.

Shannon, A. G., *Iterative Formulas Associated with Generalized Third Order Recurrence Relations,* SIAM J. Appl. Math. **23** (1972), 364–368.

Osborn, R., *A Good Generalization of the Euler-Fermat Theorem,* Math. Mag. **47** (1974), 28–31.

Lehmer, D. H., *Some Recursive Sequences,* Proc. Manitoba Conf. on Numerical Mathematics (University of Manitoba, Winnipeg, Manitoba, 1971), pp. 15–30, Department of Computer Science, University of Manitoba, Winnipeg, Manitoba, 1971.

Fermat's theorem has been generalized to matrices. See

Niven, I., *Fermat's Theorem for Matrices*, Duke Math. J. **15** (1948), 823–826.

Davis, A. S., *The Euler-Fermat Theorem for Matrices*, Duke Math. J. **18** (1951), 613–617.

Robinson, R. M., *The Converse of Fermat's Theorem*, Amer. Math. Monthly **64** (1957), 703–710.

Singmaster, D., *A Maximal Generalization of Fermat's Theorem*, Math. Mag. **39** (1966), 103–107.

See also

Laxton, R. R., *On Groups of Linear Recurrences*, I, Duke Math. J. **36** (1969), 721–736.

——, *On Groups of Linear Recurrences*. II, *Elements of Finite Order*, Pacific J. Math. **32** (1970), 173–179.

Section 5

For the law of quadratic reciprocity, see the book by Hardy and Wright cited above.

Section 6

For an introduction to algebraic number theory, see the book

Landau, E., *Grundlager de Analysis (das Rechnen mit ganzen, rationalen, irrationalen, Komplexen Zahlen) Erganzung zu den Lehrbuchen der Differential- und Intergralrechnung.*, Chelsea, New York, 1948.

A curious fact emerges. The results we quote are well known to experts and may easily be derived by them using algebraic number theory. However, they do not appear in the literature. Consequently, we cannot give a reference.

Section 7

See the paper,

Vinogradov, I. M., *The Method of Trigonometrical Sums in the Theory of Numbers,* Trav. Inst. Math. Stekloff **23** (1947), 109 (Russian).

There are three names associated with the estimation of trigonometric sums, Vinogradov, Van der Corput, and Weyl. What method to use, and what interval to use the method over, depends upon trial and error.

These are scheduling problems. This means that dynamic programming and functional equations can be used. Dynamic programming is a theory of multi-stage decision processes; see

Bellman, R., *Introduction to the Mathematical Theory of Control Processes,* Vol. I, Academic Press, New York, 1968.

——, *Introduction to the Mathematical Theory of Control Processes, Vol.* II, Academic Press, New York, 1971.

4. The Γ Function

1. Introduction

In this chapter, we want to study the Euler Γ-function.

This is one of the important functions of analysis. However, we shall not study it for itself. We are primarily interested in its application to the Riemann zeta function.

We shall begin by establishing the fundamental functional equation. Then we shall provide an asymptotic estimate using the method of Laplace. Following that we shall establish some formulas using a representation due to Weierstrass.

We shall mention some important extensions due to Siegel and Ingham.

We have merely sketched the principle result. The reader is urged to fill in the details.

2. The Basic Functional Equation

The Γ function of Euler is defined by the equation below for $\mathrm{R}\,e(s) > 0$

(1)
$$\Gamma(s) = \int_0^\infty x^{s-1} e^{-x} dx.$$

Richard Bellman, Analytic Number Theory: An Introduction   ISBN 0-8053-0452-5

An easy integration by parts yields the fundamental functional equation

(2) $\Gamma(s + 1) = s\Gamma(s)$.

By means of this equation, we can easily obtain the analytic continuation throughout the entire complex plane. Furthermore, we obtain the fact that the function defined this way has simple poles and we can obtain the residue at each pole.

The same integration by parts that yielded the fundamental functional equation shows that we have

(3) $\Gamma(n + 1) = n!$

for any integer n. Thus, the Γ function may be considered as a continuous interpolation to the factorial.

Exercises

1. Show that the analytic continuation may also be obtained by writing the integral as a sum

$$\int_0^1 x^{s-1} e^{-x} dx + \int_1^\infty x^{s-1} e^{-x} dx.$$

2. Show that the quotient of any two solutions of the functional equation is periodic.

3. Show that

$$\frac{\Gamma'(s + 1)}{\Gamma(s + 1)} = \frac{\Gamma'(s)}{\Gamma(s)} + \frac{1}{s}.$$

4. Show that the integral $\int_0^\infty e^{k\sqrt{x}} x^a dx$ may be expressed in terms of the Γ function.

5. Show that the integrals

$$\int_0^\infty \frac{\cos x}{x^a} dx, \quad \int_0^\infty \frac{\sin x}{x^a} dx \quad \text{where } 0 < a < 1$$

may be expressed in terms of the Γ function by using the following contour.

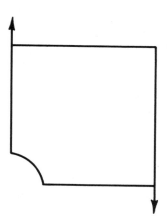

6. Show that

$$\int_0^\infty e^{-xy} x^{s-1} dx = \frac{\Gamma(s)}{y^s} \ .$$

3. Γ(s) as a Real Function

Let us now briefly consider $\Gamma(s)$ as a real function.
It is obvious by differentiation that we have

(1) $$\Gamma''(s) = \int_0^\infty (\log x)^2 x^{s-1} e^{-x} dx.$$

From this, it follows that $\Gamma(s)$ is a convex function for real s.
In the next section, we shall obtain the asymptotic behavior.

Exercises

1. Determine where $\Gamma(s)$ has its minimum value.

2. Show that the only logarithmically convex solution of the functional equation is a multiple of the Γ function. (Artin)

4. Laplace's Asymptotic Evaluation of an Integral

In this section, we follow a method of Laplace to obtain the asymptotic evaluation of $\Gamma(s)$.

Let us recall a result due to Stirling.

(1) $$n! \sim n^n e^{-n} \sqrt{2\pi n}.$$

This formula is very important in probability theory. To obtain this result and its generalization to the Γ function we use a method due to Laplace.

Let us write the integral as

(2) $$\Gamma(n+s) = \int_0^\infty x^{s-1} x^n e^{-x} dx.$$

If we now make the change of variable $x = ny$, this has the effect of making the integral

(3) $$n^{n+1} \int_0^\infty x^{s-1} e^{-n(\log x - x)} dx.$$

This change of variable has the advantage of localizing the contribution of the integral at $x = 1$. We observe that the function $\log x - x$ has its maximum value at $x = 1$. Consequently, we expect that as n becomes larger, the contribution of the integral will center about $x = 1$. It remains to make the argument rigorous. To do this we write the integral as the sum of three parts.

(4) $$\int_0^\infty = \int_0^{1-\epsilon} + \int_{1-\epsilon}^{1+\epsilon} + \int_{1+\epsilon}^\infty.$$

The choice of ϵ will be made later.

In the interval $(1 - \epsilon, 1 + \epsilon)$, we use a Taylor expansion for the function $x - \log x$. It has the form

(5) $$x - \log x = 1 + c_2 (x - 1)^2 + \cdots.$$

We know that the radius of convergence will be determined by the first singularity of the function. That occurs at $x = 0$. Consequently, if ϵ is chosen small enough, we know that the series converges. However, we do not want the full series. Rather, we want the first two terms and an error term. We know from exercises in Chap. 1 that we have

(6) $$x - \log x = 1 + c_2 (x - 1)^2 + 0(|x - 1|)^3.$$

Consequently, we have

(7) $\quad \int_0^\infty x^{s-1} e^{-n(x - \log x)} dx = \int_{1-\epsilon}^{1+\epsilon} x^{s-1} e^{-n[1+c_2(x-1)^2+0(x-1)^3]} + R_n.$

We leave it to the reader to show that R_n is of small order as n gets large.

It remains to choose ϵ. We want to choose ϵ so that R_n is small and the contribution from the $(x - 1)^3$ is small. We can do both. A suitable choice of ϵ, for example, is $1/\sqrt{n}$. We can write

(8) $\qquad\qquad\qquad e^{-n(1-x)^3} = 1 + 0(n(1 - x)^3)$

in the interval $(1 - \epsilon, 1 + \epsilon)$, where the 0-term is uniformly bounded by $2n (1 - x)^3$.

The trick is now to expand the interval $(1 - \epsilon, 1 + \epsilon)$ to the full interval $(0, \infty)$. We thus find that the asymptotic evaluation depends upon the integral

(9) $\qquad\qquad\qquad\qquad \int_0^\infty e^{-x^2} dx$

which we have evaluated in Chap. 1. This explains the appearance of the constant $\sqrt\pi$.

We thus obtain the result

(10) $\qquad\qquad\qquad \Gamma(n + s) \sim n^{s-1} n^n e^{-n} \sqrt{2\pi n},$

an interesting generalization of Stirling's formula.

Exercises

1. Show that the Γ function cannot have any zeros for any real value.

2. Obtain the asymptotic evaluation by using the functional equation for the derivative of the Γ function and Euler's constant.

5. The Representation of Weierstrass

Let us now give an important representation of Weierstrass for the Γ function. We will use this representation to obtain many other results.

This representation can be obtained from general theory. However, we will use the asymptotic evaluation. The basic idea is quite simple. We shall iterate the functional equation, use the asymptotic evaluation, and then group the terms suitably.

Let us now go through this procedure. We have

(1) $$\frac{1}{\Gamma(s)} = \frac{s}{\Gamma(s+1)} = \frac{s(s+1)\cdots(s+n)}{\Gamma(s+n+1)}$$

upon iteration. We can write this

(2) $$\frac{1}{\Gamma(s)} = \frac{s\left(1+\dfrac{s}{1}\right)\left(1+\dfrac{s}{2}\right)\cdots\left(1+\dfrac{s}{n}\right)\Gamma(n+1)}{\Gamma(s+n+1)}.$$

Unfortunately, the infinite product

(3) $$\prod^{\infty}\left(1+\frac{s}{n}\right)$$

diverges. Consequently, we introduce the convergence factor $e^{-s/n}$ and write the product in the form

(4) $$\prod_{n=1}^{\infty}\left(1+\frac{s}{n}\right)e^{-s/n}.$$

This infinite product converges. Using the Euler evaluation,

(5) $$\sum_{k=1}^{n}\frac{1}{k} = \log n + \gamma + 0\left(\frac{1}{n}\right)$$

and the asymptotic form of the Γ function we obtain the representation of Weierstrass

(6) $$\frac{1}{\Gamma(s)} = s\,e^{\gamma s}\prod_{n=1}^{\infty}\left(1+\frac{s}{n}\right)e^{-s/n}$$

Exercise

1. Show that we need only the asymptotic behavior of $\Gamma(n)/\Gamma(n+s)$.

6. Half the Sine Function

Using the Weierstrass representation, we have

(1)
$$\frac{1}{\Gamma(s)\Gamma(-s)} = s^2 \prod_{n=1}^{\infty}\left(1 - \frac{s^2}{n^2}\right)$$

$$= -s\,\frac{\sin \pi s}{\pi}.$$

Since the infinite product converges absolutely and uniformly, we can group the terms conveniently.

Using the functional equation we obtain finally the formula

(2)
$$\Gamma(s)\Gamma(1 - s) = \frac{\pi}{\sin \pi s}.$$

Exercise

1. Show that the foregoing result yields an evaluation of $\Gamma(\tfrac{1}{2})$.

7. The Duplication Formula

Consider the function

(1)
$$2^{2s-1}\left(\Gamma(s)\Gamma\left(s + \frac{1}{2}\right)\Big/\Gamma(2s)\right).$$

Using the functional equation, we see that this function is periodic of period ½. Using the asymptotic evaluation, we see that this periodic function is actually a constant. We leave it for the reader to determine the constant.

There are many ways of establishing this formula. Below we shall establish it in the manner of Euler using an infinite integral. It may also be established by using the results of a theory of a complex variable.

We also leave it to the reader to use the Weierstrass representation to obtain this formula.

Exercises

1. Show that

$$\frac{\Gamma(s)\Gamma\left(s + \frac{1}{n}\right)\cdots\Gamma\left(s + \frac{n-1}{n}\right)}{\Gamma(ns)}$$

is a simple function, and evaluate this function. (Gauss)

2. Obtain the duplication formula by using the Hecke integral and the Mellin transform.

3. Use the result of Gauss and the Mellin transform to evaluate the integral

$$\int_0^\infty \int_0^\infty \frac{e^{-\left(x_1 + x_2 + \frac{y}{x_1 x_2}\right)}}{x_1^a x_2^b} \, dx_1 \, dx_2,$$

for suitable values of a and b.

8. The Method of Stationary Phase

In Sec. 4, we considered the asymptotic behavior as s approaches infinity along the real axis. In this section, we want to consider the asymptotic behavior as s approaches infinity along the complex axis.

The method again is simple. Using an appropriate contour, we show that the change of variable $x = iy$ is legitimate. As s approaches infinity, which means as y approaches infinity, the rapid oscillation of the function makes the integral small. It is least small in the neighborhood of a stationary point. Consequently, we now repeat the procedure of Sec. 4. We will leave all the details to the reader.

9. The Beta Function

Let us write

(1) $$B(m, n) = \int_0^1 x^{m-1} (1 - x)^{n-1} \, dx.$$

This integral, the Beta function of Euler, is defined for

(2) $$\operatorname{Re}(m) > 0, \quad \operatorname{Re}(n) > 0.$$

It occurs in many parts of analysis. As we shall show, it can be easily evaluated in terms of the gamma function.

We begin with the change of variable $y = x/(1 - x)$.

This transformation has the effect of mapping the interval $(0, 1)$ into the $(0, \infty)$ in a 1-1 fashion.

We then have to evaluate the integral

(3) $$\int_0^\infty \frac{y^{m-1}}{(1 + y)^{m+n}} \, dy.$$

To evaluate this integral, we use the representation

(4)
$$\frac{1}{(1+y)^{m+n}} = \int_0^\infty e^{-(1+y)x} x^{m+n-1} dx$$

Using this representation, and inverting the order of integration, we obtain the desired result

(5)
$$B(m,n) = \frac{\Gamma(m)\Gamma(n)}{\Gamma(m+n)}$$

Exercises

1. Using the evaluation of the Beta function and the change of variable $nx = y$, obtain the asymptotic expression

$$\Gamma(n)/\Gamma(n+s) \sim n^{-s}.$$

2. Evaluate the integral $\int_0^1 dx/\sqrt{x(1-x)}$ by means of integration and the Beta function and thus evaluate $\Gamma(\frac{1}{2})$.

3. Show that the integrals $\int_0^\pi (\sin\theta)^n (\cos\theta)^n \, d\theta$ can be evaluated in terms of the Beta function.

4. Show that the $\int\int x^m y^n dx dy$ over the region $x^2 + y^2 \leqslant 1$ can be expressed in terms of the beta function. (Liouville)

5. Find the area of the ellipse $x^2/a^2 + y^2/b^2 = 1$.

6. Find the area of the circle and thus evaluate $\Gamma(\frac{1}{2})$.

7. Find the volume of the ellipsoid $x^2/a^2 + y^2/b^2 + z^2/c^2 = 1$.

10. The Integral of Siegel

An extensive generalization of the Euler integral exists,

(1)
$$\int_{A>0} |A|^{s-1} e^{-\mathrm{tr}(AB)} dA = (\cdots).$$

Let us explain the notation. The integral is over positive definite matrices. The notation $|A|$ indicates the determinant of A. The expression $\mathrm{tr}(AB)$, is the sum of the elements along the main diagonal.

The evaluation of this integral can be carried out inductively. In the bibliography and comments, we give references to some generalizations of this integral, which forms an essential part of Siegel's theory of matrix automorphic functions. An equivalent integral was found by Ingham in connection with mathematical statistics.

Miscellaneous Exercises

1. Show that $\Gamma(s)$ satisfies no polynomial equation.

2. Show that $\Gamma(s)$ satisfies no polynomial differential equation. (Holder)

3. Prove that $\Gamma(x)$ satisfies the recurrence relation

$$\Gamma'(x + 2) = (2x - 1)\Gamma'(x + 1) - x\Gamma'(x).$$

4. By considering the quotient $\Gamma'(x)/\Gamma(x + 1)$ obtain a formal continued fraction.

5. In this way obtain a continued fraction for c.

6. Does this continued fraction converge? (For questions of this type, see

Perron, O., *Die Lehre von den Kettenbrüchen,* Chelsea, New York, 1950.

Bibliography and Comments

Section 1

For more information about the gamma function see the book

Whittaker, E. T. and G. N. Watson, *A Course of Modern Analysis,* 4th ed., Cambridge University Press, Cambridge, 1935.

Section 8

An extensive account of the method of stationary phase will be found in the books

Szego, G., *Orthogonal Polynomials,* Amer. Math. Soc., Providence, Rhode Island, 1975.

Watson, G., *Bessel Functions,* Cambridge University Press, New York: Macmillan, New York, 1944.

Section 10

For the Siegel formula and the theory of automorphic matrix functions, see his book

Siegel, C. L., *The Analytical Theory of Quadratic Forms,* Mimeographed Notes, Princeton, 1934–1935.

The result of Ingham is contained in

Ingham, A. E., *An Integral which Occurs in Statistics,* Proc. Cambridge Philos. Soc. **29** (1933), 271–276.

The generalizations will be found in,

Bellman, R., *A Generalization of Some Integral Identities due to Ingham and Siegel,* Duke Math. J. **24** (1956), 571–578.

Olkin, I., *A Class of Integral Identities with Matrix Argument,* Duke Math. J. **26** (1959), 207–213.

5. Riemann Zeta Function

1. Introduction

In this chapter, we will study some properties of the Riemann zeta function. This is a very interesting function and well merits study for itself. However, we are primarily interested in it as a generating function for the elementary arithmetic functions.

In the bibliography at the end of the chapter, we will give references where further information can be obtained.

2. Dirichlet Series

The Riemann zeta function is defined by the series.

$$(1) \qquad \zeta(s) = \sum_{n=1}^{\infty} n^{-s}, \, \mathrm{Re}(s) > 1.$$

This series is the most famous example of a class of infinite series which are very important in analytic number theory, the Dirichlet series. The general form of a series of this type is

$$(2) \qquad f(s) = \sum_{n=1}^{\infty} a_n n^{-s}.$$

It is clear that the sum of two series of this form is again a series of this form.

Richard Bellman, Analytic Number Theory: An Introduction   ISBN 0-8053-0452-5

What is important is that the product of two series of this form is again a series of this form. We have

(3)
$$f(s)g(s) = \sum_{n=1}^{\infty} c_n n^{-s}$$

where

(4)
$$c_n = \sum_{kl=n} a_k b_l$$

$$= \sum_{k/n} a_k b_{n/k}.$$

Exercises

1. Prove that

$$\left(1 - \frac{1}{2^{s-1}}\right) \zeta(s) = \sum_{n=1}^{\infty} \frac{(-1)^n}{n^s}.$$

2. Prove that the series on the right converges for $\text{Re}(s) > 0$.

3. The Euler Product

Let us now give the most important result concerning the zeta function. We begin with the observation that we have the unique factorization theorem. The analytic equivalent of this is the famous representation of Euler

(1)
$$\sum_{n=1}^{\infty} \frac{1}{n^s} = \prod_{p} \left(1 - \frac{1}{p^s}\right)^{-1}.$$

What is remarkable about this result is on the left we have summation over the integers, while on the right we have a product over the primes. This formula is the basis for all the work on the asymptotic number of primes.

Exercises

1. Obtain the Dirichlet series for $\log \zeta(s)$.

2. Obtain the Dirichlet series for $\zeta'(s)/\zeta(s)$.

4. The Möbius Function

In general, it is quite difficult to find a reciprocal of a Dirichlet series. If, however, the Dirichlet series has an Euler product it is quite simple. Thus, we have

$$(1) \qquad \zeta(s)^{-1} = \sum_{n=1}^{\infty} \frac{\mu(n)}{n^s} = \prod_{p} \left(1 - \frac{1}{p^s}\right).$$

The function $\mu(n)$ is one of the most interesting functions of number theory. As we shall see, its behavior is intimately connected with the location of the complex zeros, of the zeta function.

The function $\mu(n)$ can be defined from this representation, or independently as

$$\begin{aligned} \mu(n) &= \ 1, \quad \text{if } n \text{ has an even number of simple prime factors,} \\ (2) \qquad &= -1, \quad \text{if } n \text{ has an odd number of simple prime factors,} \\ &= \ 0, \quad \text{otherwise.} \end{aligned}$$

If we use the obvious result

$$(3) \qquad \zeta(s)\zeta(s)^{-1} = 1,$$

we have the fundamental relation

$$(4) \qquad \sum_{k/n} \mu(k) \ = 0, \ \ n \neq 1,$$

$$= 1, \ \ n = 1.$$

Exercise

1. Regarding $\zeta'(s)/\zeta(s)$ as $\zeta'(s)$ times $1/\zeta(s)$, obtain an identity for the coefficient of $\zeta'(s)/\zeta(s)$.

5. The Möbius Inversion Formula

Using the properties of the Möbius function we obtain an interesting inversion formula.

We have

$$f(x) = \sum_{n=1}^{\infty} g(nx),$$

(1)

$$g(x) = \sum_{n=1}^{\infty} \mu(n)f(nx).$$

This formula may be established under various conditions. Formerly, it follows from the properties of the Möbius function by rearrangement.

6. The Squarefree Function

If we square the mobius function, we obtain a very interesting counting function. We have

(1)
$$\mu^2(n) = 1, \quad \text{if } n \text{ is squarefree,}$$
$$= 0, \quad \text{otherwise.}$$

We shall study this function in more detail in Chap. 10.

7. Multiplicative Functions

It turns out that the elementary functions have a very important property. We introduce the concept of multiplicative functions. If

(1) $$f(mn) = f(m)f(n) \quad \text{if } (m, n) = 1,$$

we say that the function $f(n)$ is multiplicative.
   If

(2) $$f(mn) = f(m) + f(n) \quad \text{if } (m, n) = 1,$$

we say that the function $f(n)$ is additive.
   We have

(3) $$f(n) = \prod_{p} f(p^{\alpha})$$

if

(4) $$n = \prod_{p} p^{\alpha}.$$

It follows that we have a generalized Euler product.

$$(5) \qquad \sum_n \frac{f(n)}{n^s} = \prod_p \left(1 + \frac{f(p)}{p^s} + \cdots\right).$$

Exercises

1. Show that $f(n^k)$ is multiplicative if $f(n)$ is.

2. Show that if $f(n)$ is additive and monotone, it must be a multiple of the logarithm. (Erdos)

3. Show that if $u(x)$ satisfies the equation $u(x + y) = u(x) + u(y)$, and is differentiable, it is of the form $u(x) = kx$. (Cauchy)

4. Show that the same conclusion holds if we merely assume that $u(x)$ is continuous.

5. Show that the same conclusion holds if we merely assume that $u(x)$ is integrable.

6. Show that if we have the inequality $|u(x + y) - u(x) - u(y)| \leqslant \epsilon$, then a constant k exists such that $|u(x) - kx| \leqslant 3\epsilon$.

7. Show that the constant 3 cannot be improved.

8. Show that if $u_n \geqslant 0$ and $u_{m+n} \leqslant u_m + u_n$ then u_n/n has a limit. (Fekete)

9. If $f_{n+1} \geqslant f_n - a_n$ where $\sum_{n=1}^{\infty} a_n$ is convergent, then f_n converges.

10. If f and g are multiplicative then the convolution of f and g is also.

8. Analytic Continuation

In this section we want to consider $\zeta(s)$ as an analytic function.

The first task is to extend $\zeta(s)$ over the whole plane. This can be done in many ways.

One way is to use the fundamental representation

$$(1) \qquad \Gamma(s)\zeta(s) = \int_0^\infty \frac{e^{-x}x^{s-1}\,dx}{1 - e^{-x}} \; .$$

We now write this integral as the sum of two parts.

$$(2) \qquad \int_0^\infty = \int_0^1 + \int_1^\infty .$$

The second term is an entire function of s. Let us then consider the first term. We have

(3)
$$\int_0^1 \frac{e^{-x} x^{s-1} dx}{1 - e^{-x}} = \int_0^1 x^{s-1} \left(\frac{e^{-x} - 1}{1 - e^{-x}} \right) dx + \int_0^1 x^{s-2} dx.$$

Consider the function

(4)
$$\frac{e^{-x}}{1 - e^{-x}} - \frac{1}{x}.$$

This function has its nearest singularity at $2\pi i$. Consequently, the power series converges for $|x| < 2\pi$.

Integrating term by term, we obtain the expansion

(5)
$$\Gamma(s)\zeta(s) = \sum_{n=0}^{\infty} \frac{c_n}{s+n} + \frac{1}{s-1}.$$

The last term comes from integration. This furnishes the desired analytic continuation. The singularities at $-n$ are cancelled by the zeros of $1/\Gamma(s)$.

Exercises

1. Show that

$$\zeta(s) = \int_0^\infty \frac{x - [x]}{x^s} dx.$$

2. Show that

$$\lim_{n \to \infty} \left(\sum_{n=1}^N \frac{1}{n^s} - \frac{N^{1-s}}{1-s} \right) = \zeta(s),$$

for $0 < \mathrm{Re}(s) < 1$.

3. Show that the formula

$$\left(1 - \frac{1}{2^{s-1}} \right) \zeta(s) = \sum_{n=1}^{\infty} \frac{(-1)^n}{n^s}.$$

4. Show that this formula gives the singularity at $s = 1$. Show that the other singularities do not exist.

9. The Functional Equation

Let us now derive the famous functional equation for $\zeta(s)$.

We shall employ the well-known functional equation for a theta function. We shall derive that equation in Chap. 6 and again in Chap. 7. Here we shall merely use it without proof.

We use the representation

(1)
$$\Gamma\left(\frac{s}{2}\right)\zeta(s) = \int_0^\infty t^{s/2-1} \sum_{n=1}^\infty e^{-n^2 t} dt$$

We write

(2)
$$\int_0^\infty = \int_0^1 + \int_1^\infty.$$

Next, we write

(3)
$$\sum e^{-n^2 t} = \frac{1}{2} \sum_{n=-\infty}^\infty e^{-n^2 t}.$$

In the interval $(1, \infty)$, we make the change of variable $t = 1/r$, and use the transformation formula for the theta function.

(4)
$$\sum_{-\infty}^\infty e^{-n^2 t} = \sqrt{\frac{\pi}{t}} \sum_{n=-\infty}^\infty e^{-n^2 \frac{\pi}{t}}.$$

The final result is

(5)
$$\pi^s \Gamma\left(\frac{s}{2}\right)\zeta(s) = \pi^{(1-s)}\Gamma\left(\frac{1-s}{2}\right)\zeta(1-s).$$

This shows that $\pi^s\Gamma(s/2)\zeta(s)$ is symmetrical about the line $\frac{1}{2} + it$.

Exercises

1. Let $r(n)$ denote the number of representations of n as a sum of two squares.

Establish a functional equation for

$$\sum_{n=1}^{\infty} \frac{r(n)}{n^s} \; .$$

2. Let $r_k(n)$ denote the number of representations of n as the sum of k^2. Establish a functional equation for the associated Dirichlet series.

10. The Riemann Hypothesis

The famous Riemann hypothesis is that all the nontrivial zeros of $\zeta(s)$ are on the line $\frac{1}{2} + it$.

Despite the best efforts of many mathematicians this has defied proof. As we shall see, it would have many interesting consequences for error terms.

11. Asymptotic Behavior

Fortunately, we have no need of a good point estimate of $\zeta(s)$ for our purposes. We can use a very crude estimate. To obtain this estimate, we use the series

$$(1) \qquad\qquad \left(1 - \frac{1}{2^{s-1}}\right) \zeta(s) = \sum_{n=1}^{\infty} \frac{(-1)^{n-1}}{n^s}$$

Our procedure is very simple. We break the sum into two parts, and estimate each part crudely. To obtain the required asymptotic estimate it remains to choose the way we decompose the sum. Let us write

$$(2) \qquad\qquad \sum_{n=1}^{\infty} = \sum_{n=1}^{N} + \sum_{N+1}^{\infty} .$$

Choosing N to be $[t]$, we see we have the estimate

$$(3) \qquad\qquad |\zeta(\tfrac{1}{2} + it)| = 0(t^{\frac{1}{2}}).$$

We can obtain a better point estimate using a mean value estimate. Let us use the result

$$(4) \qquad\qquad \int_1^T |\zeta(\tfrac{1}{2} + it)|^2 \, dt = 0(T \log T).$$

We shall obtain a more precise result in Chap. 12. As the exercises show, this result can be obtained in various ways.

To obtain the point estimate, we use the fact that $\zeta(s)$ is an analytic function of s. Thus, we can employ the representation of Cauchy.

$$(5) \qquad \zeta(s) = \frac{1}{2\pi i} \int_R \frac{\zeta(w)dw}{w-s}.$$

It remains to choose the region R. We choose the following region

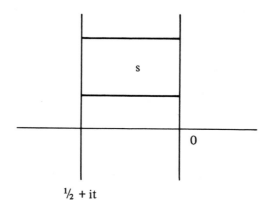

$\tfrac{1}{2} + it$

Using this contour, we can estimate $\zeta(s)$ in terms of various mean values. To estimate the integral, we use the Cauchy-Schwarz inequality. We have

$$(6) \qquad \left| \int \frac{\zeta(w)}{s-w}dw \right| \leqslant \left(\frac{|\zeta(w)|^2}{|s-w|}dt \right)^{1/2} \left(\int \frac{dw}{|s-w|} \right)^{1/2}$$

The contribution from the other 3 sides of the rectangle are of lower order. The final result is

$$(7) \qquad |\zeta(\sigma + it)| = 0(\log t^{3/2})$$

for $\sigma > \tfrac{1}{2}$.

Exercises

1. Use the equation in (1) to establish the mean value result stated.

2. Estimate in a similar fashion the integral of $\zeta(s)$.

3. Estimate in a similar fashion the derivative of $\zeta(s)$.

4. What contours do we use if we want to estimate the function on the critical line.

12. Uses of the ζ Function

Once we have established the analytic properties of $\zeta(s)$ we can use it to obtain the behavior of many familiar functions of analysis.

The basic idea is quite simple. We use the Mellin transform to obtain a representation, use the Mellin inversion formula, and then shift the contour.

Let us give an example. We have

$$(1) \qquad \int_0^\infty t^{s-1} \sum_{n=1}^\infty \frac{e^{-nt}}{n^a} \, dt = \Gamma(s)\zeta(s+a).$$

From this follows

$$(2) \qquad \sum_{n=1}^\infty \frac{e^{-nt}}{n^a} = \frac{1}{2\pi i} \int_C t^{-s} \Gamma(s)\zeta(s+a) \, ds.$$

If we shift the contour to the left, we obtain the representation

$$(3) \qquad \sum_{n=1}^\infty \frac{e^{-nt}}{n^a} = \frac{\Gamma(1-a)}{t^{1-a}} \cdot \zeta(a) + \frac{1}{2\pi i} \int_{C'} t^{-s} \Gamma(s)\zeta(s+a) \, ds.$$

Here C' is a contour which is a straight line to the left of $s = 0$.

Exercises

1. Find the behavior as $t \to 0$ of $\sum_{n=1}^\infty e^{-n^k t}$.

2. Show that

$$\sum_{n=1}^\infty \log(1 - e^{-nt}) = \sum_{m,n} \frac{e^{-mnt}}{m} = \sum_{n=1}^\infty \frac{1}{n} \frac{e^{-nt}}{1 - e^{-nt}}.$$

3. Hence, show that

$$\prod_{n=1}^{\infty}(1 - e^{-nt}) = e^{f(t)}$$

where

$$f(t) = \sum_{n=1}^{\infty}\left(\sum_{k\mid n}\frac{1}{k}\right)e^{-nt}.$$

4. Hence, determine the asymptotic behavior as $t \to 0$ of $\prod_{n=1}^{\infty}(1 - e^{-nt})$.

5. Determine the asymptotic behavior as $x \to 1$ of $\sum_{n=1}^{\infty}x^{n}/(1 - x^{n})$.

6. Determine the asymptotic behavior as x and $y \to 0$ of $\prod_{m,n=1}^{\infty}(1 - e^{-(mx+ny)})$.

13. Mean Values of the ζ Function

We can also use the foregoing representation to obtain mean values. Remembering the dictum of Jacobi, we can use the foregoing representation of $\zeta(s)$. We have

(1)
$$\zeta(\alpha) = \sum_{n=1}^{\infty}\frac{e^{-nx}}{m^{\alpha}} + \frac{\Gamma(1 - \alpha)}{x^{+1-\alpha}} + 0(x^{\sigma})$$

where $\frac{1}{2} < \alpha < 1, \quad \sigma > 0$.

14. The Prime Number Theorem

Let us say a few words about the prime number theorem.

Let us denote by $\pi(x)$ the number of primes less than or equal to x. Analytically, we have

(1)
$$\pi(x) = \sum_{p \leqslant x} 1.$$

Tables showed that

(2)
$$\pi(x) \sim \frac{x}{\log x}.$$

Finally, in 1899, Hadamard and de la Vallee Poussin established this result rigorously. Their proof, which is classical, depended upon showing that the zeta function did not vanish on the line $\sigma = 1$, and depended heavily upon the theory of a complex variable.

Miscellaneous Exercises

1. Write aXb to denote the convolution product. Show that

$$aXb = bXa$$

2. Show that $aXbXc$ is unambiguously defined.

3. Show that

$$\zeta(s)f(s) = \sum b_n/n^s$$

where

$$b_n = \sum_{k|n} a_k.$$

4. Show that

$$\zeta^2(s) = \sum d_n/n^s$$

where

$$d(n) = \sum_{k|n} 1.$$

5. Prove that

$$\prod_{p \leqslant N} \left(1 - \frac{1}{p}\right)^{-1} > \sum_{n=1}^{N} \frac{1}{n}.$$

6. Hence, show that

$$\sum_{p \leqslant N} \frac{1}{p} \geqslant \log\log N - c_1.$$

7. Show that

$$\log \zeta(s) + c_2 \geqslant \sum_{p \leqslant N} \frac{1}{p^s}$$

for any s greater than 1.

8. Choose

$$s = 1 + \frac{1}{(\log N)(\log \log N)}$$

and hence show that we have

$$(1 + o(1))\log \log N \geqslant \sum_{p \leqslant N} \frac{1}{p} - c_3.$$

9. Hence show that

$$\sum_{p \leqslant N} \frac{1}{p} \sim \log \log N.$$

10. Obtain conditions in $f(n)$ so that we have a similar result for

$$\sum_{p \leqslant N} \frac{f(p)}{p}.$$

11. We say the two integers m and n are relatively disjoint if they have no power of two in common in their representation. We shall use the notation $[m, n] = 1$. We say that a function is additively disjoint if it satisfies the functional equation

$$f(m + n) = f(m) + f(n, [m,n]) = 1.$$

Construct a theory of additively disjoint functions.

12. Construct a theory of multiplicative functions for Gaussian integers.

13. Is there an analog of the fundamental theorem of arithmetic for matrices? (Cauchy's theorem was extended to matrices by Poincaré, H., *Sur les groupes continus,* Trans. Cambridge Philos. Soc. **18** (1899), 220-225. See the Bellman book *Introduction to Matrix Analysis,* McGraw-Hill, New York, 1960; 2nd ed., 1970.)

14. Is there a theory of multiplicative functions for matrices?

15. Is there an analog of the fundamental theorem of arithmetic for quaternions? (A theory of quaternion functions similar to the theory of functions of a complex variable was constructed by Fueter. See Fueter, R., *Functions of a Hyper Complex Variable,* University of Zurich, 1948-1949, reprinted by Argonne National Laboratory, 1959.)

16. Is there a theory of multiplicative functions for quarternions?

17. Does the serie:

$$\sum_{n=1}^{\infty} \frac{1}{n^{1+it}}$$

converge for t not equal to 0?

18. Consider the functional equation

$$Y(s + t) = Y(s)Y(t), \qquad Y(0) = I.$$

Show that if we assume that Y is differentiable, Y must have the form $Y(t) = e^{At}$. Here Y and A are matrices.

19. Show that the same conclusion holds if we assume that Y is continuous. (Polya) (See the book

Bellman, R., *Introduction to Matrix Analysis,* McGraw-Hill, New York, 1960; 2nd ed., 1970.)

20. By using the relation $\int_R \zeta(s)\, ds = 0$, where R is a suitably chosen rectangle, derive a nontrivial estimate for $\int_0^T \zeta(\frac{1}{2} + it)\, dt$.

21. We have the relation

$$\int_R \zeta(s)\zeta(1 - s)\, ds = 0.$$

Can this relation, plus the functional equation be used to obtain the mean values on the critical strip?

Bibliography and Comments

Section 1

The Riemann zeta function is a very interesting function. For further information, see the books

Whittaker, E. T. and G. N. Watson, *A Course of Modern Analysis,* 4th ed., Cambridge University Press, Cambridge, 1935.
Titchmarsh, E. C., *The Riemann Zeta Functions,* Cambridge University Press, Cambridge, 1930.

Section 2

The general theory of Dirichlet series is quite difficult. See the books,

Titchmarsh, E. C., *Theory of Functions,* Oxford University Press, 1939.
Riesz, M. and G. H. Hardy, *The General Theory of Dirichlet's Series,* Cambridge University Press, Cambridge, 1952.

For convolutions, see

Narkiewicz, W., *On a Class of Arithmetical Convolutions,* Colloq. Math. **10** (1963), 81–94.
Horadam, E. M., *A Calculus of Convolutions for Generalized Integers,* Nederl. Akad. Wetensch. Proc. Ser. A, **66**-Indag. Math. **25** (1963), 695–698.
Davison, T. M. K., *On Arithmetic Convolutions,* Canad. Math. Bull. **9** (1966), 287–296.
Lambek, J., *Arithmetical Functions and Distributivity,* Amer. Math. Monthly **73** (1966), 969–973.
Segal, *A Note on Dirichlet Convolutions,* Canad. Math. Bull. **9** (1966), 457–462.
Subbarao, M. V., *Arithmetic Functions and Distributivity,* Amer. Math. Monthly **75** (1968), 984–988.
Shapiro, H. N., *On the Convolution Ring of Arithmetic Functions,* Comm. Pure Appl. Math. **25** (1972), 287–336.
Popken, J., *An Arithmetical Property of a Class of Dirichlet's Series,* Nederl. Akad. Wetensch. Proc. **48**, pp. 517–534; Indag. Math. **7** (1945), 105–122.
Ryden, R. W., *Groups of Arithmetic Functions under Dirichlet Convolution,* Pacific J. Math. **44** (1973), 355–360.
Subbarao, M. V., *On Some Arithmetic Convolutions, The Theory of Arithmetic Functions* (Proc. Conf., Western Michigan Univ., Kalamazoo, Michigan, 1971), pp. 181–192. Lecture Notes in Math., Vol. 251, Springer, Berlin, 1972.

Section 3

The study of general Euler products was carried out by Hecke.

Section 4

See

Lehmer, D. H. and S. Selberg, *A Sum Involving the Function of Möbius,* Acta Airth. **6** (1960), 111–114.

Sastry, K. P. R., *On the Generalized Type Möbius Functions,* Math. Student **31** (1963), 85–88.

Apostol, T. M., *A Characteristic Property of the Möbius Function,* Amer. Math. Monthly **72** (1963), 279–282.

Evelyn, C. J. A., *A Relationship for the Möbius Function,* Quart. J. Math. Oxford Ser. **17** (2) (1966), 281.

Smith, D. A., *Incidence Functions as Generalized Arithmetic Functions.* I, Duke Math. J. **34** (1967), 617–633.

Albis, V. S., *A Note on Generalized Möbius μ-Functions,* Rev. Colombiana Mat. **2** (1968), 6–11.

Segal, S. L., *On Convolutions with Möbius Function,* Proc. Amer. Math. Soc. **34** (1972), 365–372.

Suryanarayana, D., *New Inversion Properties of μ and μ,* Elem. Math. **26** (1971), 136–138.

Wintner, A., *The Lebesgue Constants of Möbius' Inversion,* Duke Math. J. **11** (1944), 853–867.

Lindstrom, B., *On Möbius Functions and a Problem in Combinatorial Number Theory,* Canad. Math. Bull. **14** (1971), 513–516.

Daykin, D. E., *A Generalization of Möbius Inversion Formula,* Simon Stevin **46** (1972/73), 141–146.

Section 5

See

Satyanarayana, U. V., *On the Inversion Property of the Möbius μ-Function,* Math. Gaz. **47** (1963), 38–42.

Daykin, D. E., *Generalized Mobius Inversion Formula,* Quart. J. Math. Oxford Ser. **15** (2) (1964), 349–354.

Satyanarayana, U. V., *On the Inversion Property of the Möbius μ-Function.* II, Math. Gaz. **49** (1965), 171–178.

Chawdhury, M. R., *On the Möbius Inversion Formula,* Punjab Univ. J. Math. (Lahore) **3** (1970), 29–34.

Section 7

A great deal has been done on multiplicative functions. Some representative papers are:

Venkataraman, C. S., *The Ordinal Correspondence and Certain Classes of Mul-*

tiplicative Functions of Two Arguments, J. Indian Math. Soc. (N.S.) **10** (1946), 81-101.

——, *Classification of Multiplicative Functions of Two Arguments Based on the Identical Equation,* J. Indian Math. Soc. (N.S.), **13** (1949), 17-22.

Bell, E. T., *Solution of a Functional Equation in the Multiplicative Theory of Numbers,* Math. Mag. **24** (1951), 233-235.

Duncan, R. L., *A Class of Additive Arithmetical Functions,* Amer. Math. Monthly **69** (1962), 34-36.

Delange, H., *On a Class of Multiplicative Arithmetical Functions,* Scripta Math. **26** (1963), 121-141.

de Bruijn, N. G. and J. H. van Lint, *Incomplete Sums of Multiplicative Functions.* I, II, Nederl. Akad. Wetensch. Proc. Ser. A 67-Indag. Math. **26** (1964), 339-347, 348-359.

Goldsmith, D. L., *On the Multiplicative Properties of Arithmetic Functions,* Pacific J. Math. **27** (1968), 283-304.

Tull, J. P., *A Theorem in Asymptotic Number Theory,* J. Austral. Math. Soc. **5** (1965), 196-206.

Rearick, D., *Semi-Multiplicative Functions,* Duke Math. J. **33** (1966), 49-53.

Erdös, P. and A. Rényi, *On the Mean Value of Nonnegative Multiplicative Number-Theoretical Functions,* Michigan Math. J. **12** (1965), 321-338.

Gioia, A. A. and M. V. Subbarao, *Identities for Multiplicative Functions,* Canad. Math. Bull. **10** (1967), 65-73.

Goldsmith, D. L., *A Note on Sequences of Almost-Multiplicative Arithmetic Functions,* Rend. Mat. **3** (6) (1970), 166-170.

Halász, G., *On the Mean Value of Multiplicative Number Theoretic Functions. Number Theory* (Colloq., Janos Bolyai Math. Soc.), Debrecan, 1968, pp. 117-121, North-Holland, Amsterdam, 1970.

Chidambaraswamy, J., *On the Functional Equation $F(mn)F((m, n)) = F(m) F(n)f(m, n))$,* Portugal Math. **26** (1967), 101-107.

Wintner, A., *Mean-Values of Arithmetical Representations,* Amer. J. Math. **67** (1945), 481-485.

Scourfield, E. J., *Non-Divisibility of Some Multiplicative Functions,* Acta Arith. **22** (1972/73), 287-314.

Yocum, K. L., *Totally Multiplicative Functions in Regular Convolution Rings,* Canad. Math. Bull. **16** (1973), 119-128.

There is a strong connection between number theory and probability theory. See the papers

Rényi, A., *Probabilistic Methods in Number Theory,* Proc. Internat. Congress Math., 1958, pp. 529-539, Cambridge University Press, New York, 1960.

Kac, M., *Probability Methods in Some Problems of Analysis and Number Theory,* Bull. Amer. Math. Soc. **55** (1949), 641-665.

Erdös, P., *On Additive Arithmetical Functions and Applications of Probability* to Number Theory, Proc. Internat. Congress Math., 1954, Amsterdam, Vol. III, pp. 13-15, Noordhoff, Groningen; North-Holland, Amsterdam, 1956.

Kubilyus, I. P., Probabilistic Methods in the Theory of Numbers, Amer. Math. Soc. Transl. 19 (2) (1962), 47-85.

Kosambi, D. D., Statistical Methods in Number Theory, J. Indian Soc. Agric. Statist. 16 (1964), 126-135.

Erdös, P., On Some Applications of Probability to Analysis and Number Theory, J. London Math. Soc. 39 (1964), 692-696.

Galambos, J., Limit Distribution of Sums of (Dependent) Random Variables with Applications to Arithmetical Functions, A. Wahrscheinlichkeitstheorie und Verw. Gebiete 18 (1971), 261-270.

Section 9

A very interesting problem is whether this functional equation determines the function. As stated above, questions of this type were investigated in great detail by Hecke. See

Bochner, S., On Riemann's Functional Equation with Multiple Gamma Factors, Ann. of Math. 67 (2) (1958), 29-41.

Berndt, B. C., The Functional Equation of Some Dirichlet Series. II, Proc. Amer. Math. Soc. 31 (1972), 24-26.

This was the second method used by Riemann to derive the functional equation. See

Riemann, B., Über die Anzahl der Primzahlen unter einer Gegebenen Grösse, Werke 2, pp. 145-153, 1892. (Collected works of Bernhard Riemann, Dover, New York, 1953.)

See

Chandrasekharan, K. and H. Joris, Dirichlet Series with Functional Equations and Related Airthmetical Identities, Acta Arith. 24 (1973), 165-191.

Section 10

For a discussion of the Riemann hypothesis see the book by Titchmarsh cited above. With the aid of this hypothesis, many error terms can be improved.

Hardy showed that there were infinitely many zeros on the critical line. For his method, see the book by Whittaker and Watson cited above. A. Selberg showed that a positive fraction of all zeros were on the critical line; for his method, see the book by Titchmarsh cited above. Levinson showed that more than one third of all the zeros were on the critical line using a different approach; see the paper,

Levinson, N., "Complex Variables". San-Fransisco, Holden-day, 1970.

Hilbert is reputed to have said, "If I woke after a thousand years, the first question I would ask is, 'Has the Riemann hypothesis been settled yet?'" He also gave a talk in which he discussed three famous problems of mathematics: the Riemann hypothesis, Fermat's last theorem, the α^β hypothesis. He said that so much had been done on the theory of entire functions that it was clear that the Riemann hypothesis would be settled in about twenty years at most. He said further that so much had been done in algebraic number theory that Fermat's last theorem would also be settled soon. However, he added, the α^β hypothesis was forever beyond the power of mathematics.

The α^β hypothesis was done by Gelfand and Shneider independently in about fifteen years.

See also,

Good, I. J. and R. F. Churchhouse, *The Riemann Hypothesis and Pseudorandom Features of the Möbius Sequence,* Math. Comp. **22** (1968), 857–861.

The zeta function can be generalized in many ways. Perhaps most important is the zeta function in fields of characteristic p. This was done by Artin. By examination of particular cases, he was led to enunciate a Riemann hypothesis for these functions. This was proved in the cubic case by Hasse using elliptic functions. The general case was done by A. Weil using Abelian functions and algebraic geometry.

The result has important applications to the estimation of trigonometric sums.

Wintner, A., *Random Factorizations and Riemann's Hypothesis,* Duke Math. J. **11** (1944), 267–275.

Hejhal, D. A., *A Remark of the Lindelöf Hypothesis,* Bull. Amer. Math. Soc. **80** (1974), 695–699.

Section 11

The pointwise asymptotic behavior of $\zeta(s)$ seems to be quite difficult. At present, the best results are obtained by the use of trigonometric sums, as discussed in the paper,

Vinogradov, I. M., *The Method of Trigonometrical Sums in the Theory of Numbers,* Trav. Inst. Math. Stekloff **23** (1947), 109 (Russian).

Section 14

For an excellent account of the classical method for the proof of the prime number theorem, see the book

Chandrakersan, K., *The Elementary Arithmetic Functions,* Springer, Berlin, 1970.

6. The Poisson Summation Formula

1. Introduction

In this chapter we shall discuss one of the most important tools of analytic number theory, the Poisson summation formula. We shall give three approaches to this fundamental result. As usual, the reason why we give different approaches is that each approach generalizes in a different direction, as we shall see.

We shall discuss an approach using Fourier series. Then we shall present a method which depends upon Dirichlet series. Finally, we shall indicate how the Laplace transform may be used.

We shall indicate at various places how this formula may be applied.

2. Fourier Series

Since our first method of proof of the Poisson summation formula will be based upon the theory of Fourier series, let us briefly digress to present the rudiments of this theory. Let us repeat a little of Chap. 2.

Consider the set of exponential functions $\{e^{2\pi inx}\}$, where n assumes the values $n = 0$, $n = \pm1$, $n = \pm2$, and so on. These are clearly periodic functions of x of period one. It is plausible, from a number of physical considerations, to suspect that every continuous function of x which is periodic of period one can be represented as a linear combination of these particular functions. In other words, our physical surmise is that every periodic motion is a superposition of these simple periodic motions.

Setting

$$(1) \qquad f(x) = \sum_{n=-\infty}^{\infty} a_n e^{2\pi inx},$$

Richard Bellman, Analytic Number Theory: An Introduction   ISBN 0-8053-0452-5

how do we determine the coefficients? Fourier solved the problem by expanding both sides as power series in x and "solving" the resulting system of linear equations for the a_n. Following Euler, let us employ the orthogonality relation

(2) $$\int_0^1 e^{2\pi imx} e^{-2\pi inx} dx = \delta_{m-n}.$$

Here δ_{m-n} is the Kronecker delta symbol defined by the relations $\delta_{m-n} = 0$, $m \neq n$, $\delta_{m-n} = 1$, $m = n$. It is reasonable then to suppose that if (1) holds, then the coefficient a_n is determined by the simple formula

(3) $$a_n = \int_0^1 f(x) e^{-2\pi inx} dx.$$

Oddly, despite the physical simplicity of the situation, its mathematical aspects are both extraordinarily complex and subtle. The difficulty is essentially that the physical, intuitive concept of a continuous function is far too naive for the mathematical definition of a continuous function. Consequently, we circumvent a number of irritating pitfalls by working backwards. We start with a continuous function $f(x)$, (a type of function sufficiently general for our subsequent purposes) and form the sequence of Fourier coefficients, $\{a_n\}$, $n = 0$, $\pm 1, \pm 2, \cdots$, by means of the formula (3). Using these coefficients, we form the series

(4) $$g(x) = \sum_{n=-\infty}^{\infty} a_n e^{2\pi inx}$$

where x is a real variable, lying in the open interval $(0, 1)$. This series yields a new function $g(x)$, defined where (4) converges.

We then ask ourselves two questions:

(a) Does (4) converge for all x, and, if not, for what values of x does it converge?

(b) When (4) converges, is it equal to the function $f(x)$?

Question (a) remains unanswered up to the present day. Question (b) is answered in the affirmative. The series (4) converges to $f(x)$ if it converges at all. If we replace convergence in the usual sense by $(C, 1)$ summability, then following Fejer, we can obtain elegant and satisfying results. Fortunately, for the applications which we require, we can get by with very simple considerations.

If

(5) $$\sum_{n=-\infty}^{\infty} |a_n| < \infty,$$

the infinite series (4) converges uniformly (and of course absolutely) for all x in $(0, 1)$ to $f(x)$.

It is clear that the convergence of the series in (5) implies the uniform and absolute convergence of the infinite series (4). Hence, we have

(6)

$$\int_0^1 g(x)e^{-2\pi inx}\,dx = \int_0^1 \left(\sum_{k=-\infty}^{\infty} a_k e^{2\pi ikx} \right) e^{-2\pi inx}\,dx$$

$$= \sum_{k=-\infty}^{\infty} a_k \int_0^1 e^{2\pi ikx} e^{-2\pi inx}\,dx$$

(by virtue of uniform convergence)

$$= a_n.$$

From this, it follows that $f(x)$ and $g(x)$ are two continuous functions of x having the same Fourier coefficients; that is,

(7)

$$\int_0^1 [f(x) - g(x)]\, e^{2\pi inx}\,dx = 0,$$

for $n = 0, \pm 1, \cdots$.

It is now plausible that the function $f(x) - g(x)$ must be identically zero, and it is indeed true. Since this result can be established in a variety of ways, without further recourse to the theory of Fourier series, we shall accept it without furnishing any proof here. As usual, ingenuity is required if standard theory is not used. On the other hand, one of the basic results of the theory of Fourier series—the summability theorem of Fejer—yields the result immediately as a corollary of a far more general result.

Exercise

1. Consider the function $\sum_n e^{2\pi inx}/(n + a)^s$ where $s > 0$ and $0 < a < 1$. Obtain the Fourier series for a. (Hurwitz)

3. The Poisson Summation Formula

An important application of the theory of Fourier series is to the Poisson summation formula. This elegant and powerful technique can be used to derive a number of significant results in analysis.

Let $f(x)$ be a continuous function of x, defined for $-\infty < x < \infty$. Form the periodic function

(1)
$$g(x) = \sum_{n=-\infty}^{\infty} f(x + n).$$

For the moment, let us proceed quite formally, assuming that the series converges, and that the manipulations that follow are valid, and so on. Subsequently, we shall present a rigorous treatment.

It is clear that

(2)
$$g(x) = g(x + 1).$$

Let us now invoke a mathematical principle first explicitly enunciated, and systematically exploited, by Hecke: A periodic function should always be expanded in a Fourier series.

To obtain the Fourier coefficients of $g(x)$ we write

$$a_k = \int_0^1 g(x)e^{-2\pi ikx}dx = \int_0^1 \left[\sum_{n=-\infty}^{\infty} f(x + n) \right] e^{-2\pi ikx}dx$$

(3)
$$= \sum_{n=-\infty}^{\infty} \int_0^1 f(x + n)e^{-2\pi ikx}dx$$

$$= \sum_{n=-\infty}^{\infty} \int_n^{n+1} f(x)e^{-2\pi ikx}dx = \int_{-\infty}^{\infty} f(x)e^{-2\pi ikx}dx.$$

Consequently, provided we can justify all of the above, we have the identity

(4)
$$\sum_{n=-\infty}^{\infty} f(x + n) = \sum_{k=-\infty}^{\infty} e^{2\pi ikx} \int_{-\infty}^{\infty} f(x_1)e^{-2\pi ikx_1}dx_1.$$

The case $x = 0$ yields the Poisson summation formula:

(5)
$$\sum_{n=-\infty}^{\infty} f(n) = \sum_{k=-\infty}^{\infty} \int_{-\infty}^{\infty} f(x_1)e^{-2\pi ikx_1}dx_1.$$

Exercise

1. Evaluate

$$\int_0^n (n - [x])f'(x)\,dx,$$

use the Fourier series for $(x - [x])$, and thus obtain another proof of the Poisson summation formula.

4. Some Simple Sufficient Conditions

Examining the procedure of Sec. 3, we see that the imposition of some very simple conditions, which are easy to apply, will justify our procedures. Let us suppose that
(a) the function $f(x)$ is continuous for all real finite x;
(b) the infinite series,

$$\sum_{n=-\infty}^{\infty} f(x + n)$$

converges uniformly in every finite x interval;
(c) the infinite integral $\int_{-\infty}^{\infty} |f(x)|\,dx$ converges;
(d) the series

$$\sum_{k=-\infty}^{\infty} |a_k|,$$

converges, where a_k is the Fourier coefficient determined in 3.3.
   Then, under the foregoing conditions, the two sides of 3.4 exist and are equal for all x. This is Section 3 of this chapter.

5. Application to the Theta Function

Probably, the most famous application of the Poisson summation formula is to obtain the modular transformation. If we consider the sum $\Sigma_n\, e^{-n^2 \pi t}$, we see that it fulfills all the above conditions. Furthermore, the integral

(1)
$$\int_{-\infty}^{\infty} e^{-y^2 \pi t}\,dy$$

can be evaluated, as we know. Consequently, we obtain the celebrated modular transformation.

Exercises

1. Apply the Poisson summation formula to the sum $\Sigma_n \ e^{-n^2\pi t + 2n\pi iz}$ and thus obtain the transformation for the θ function.

2. Using the Laplace transform and the Hecke integral, derive from this formula the Lipshitz formula.

3. From the Lipshitz formula, derive the functional equation for the θ function.

6. Multi-dimensional Version

There is no difficulty in obtaining a multi-dimensional version of the Poisson summation formula. All that we need is the multi-dimensional analog of Fourier series. As before, we can give simple sufficient conditions for the validity of this formula.

7. Interesting Domains

In the application of the multi-dimensional formula it is important to have various domains of interest in analysis. In analytic number theory, we find the domain of Hecke, suggested by algebraic number theory, and the domain of Siegel, suggested by matrix theory. In analysis, we have the domain of Bochner,

$$x_1 \geqslant \sqrt{x_2^2 + \cdots + x_n^2}\,.$$

As we have shown in Chap. 1, other interesting domains arise naturally.

8. The Evaluation of $\int_{-\infty}^{\infty} e^{-(x,\,Ax)}dx$

For the application of the multi-dimensional Poisson formula to the multi-dimensional modular transformation, we must evaluate the $\int_{-\infty}^{\infty} e^{-(x,\,Ax)}dx$. To accomplish this, it is convenient to use a little matrix analysis. As we indicate in the exercises, this is not necessary.

We have

(1)
$$A = T^{-1} \begin{pmatrix} \lambda_1 & & & 0 \\ & \lambda_2 & & \\ & & \ddots & \\ 0 & & & \lambda_N \end{pmatrix} T,$$

where $\lambda_1 \cdots \lambda_N$ are the characteristic roots.

Let us in the integral make the transformation $x = Ty$. Using the foregoing, the integral becomes

(2)
$$\int_{-\infty}^{\infty} e^{-(\lambda_1 y_1^2 + \cdots + \lambda_N y_N^2)} dy.$$

Using the one-dimensional case, we have the result

(3)
$$\int_{-\infty}^{\infty} e^{-(y, Ay)} dy = \pi^{N/2}/|A|^{1/2}$$

where $|A|$ represents the determinant of A.

Exercises

1. Show that the above formula is valid if A has the form $B + iC$ where B is positive definite and C is symmetric and commutes with B.

2. Show that the above formula is valid if B is positive definite and C is symmetric.

3. Prove the formula in the text by reducing the quadratic form to a sum of squares. (Lagrange)

9. The General Modular Transformation

Using the multi-dimensional Poisson summation formula and the integral above, we can readily obtain the general modular transformation.

What is difficult is in any particular case to pick out the theta functions that are pertinent. Here we are guided by algebraic number theory or by matrix theory in many cases.

10. Sums of Squares

We easily see that

(1)
$$\left(\sum_n e^{-n^2 z}\right)^k = \sum_{n=1}^{\infty} r_k(n) e^{-nt},$$

where $r_k(n)$ is the number of representation of n as a sum of k squares. If we make the change of variable $e^{-t} = z$, we can use Cauchy's formula to express

$r_k(n)$. We expect that the modular transformation will help us obtain an expression for $r_k(n)$. To accomplish this, the Farey disection is very useful.

This program was carried out by Hardy. In the general case, the same method can be used for Waring's problem, as was done by Hardy and Littlewood. A different procedure, based on trigonometric sums, was used by Vinogradov.

11. Partition Functions

Another important problem in number theory is the representation of an integer by integers of a given set. The most important example of this is the representation of an integer as a sum of integers. This is the famous partition problem, which has been studied extensively.

A modular function again plays a central role, the modular function of Dedekind.

12. Gaussian Sums

One of the beautiful applications of the transformation formula of the theta function is to the evaluation of the Gauss sum,

$$(1) \qquad S(p, q) = \sum_{r=0}^{q-1} e^{-\pi i r^2 p/q},$$

where p and q are relatively prime integers. These sums are of great importance in number theory.

To find the connection between this sum and the theta functions, let us take the function

$$(2) \qquad f(t) = \sum_{n=-\infty}^{\infty} e^{-n^2 t} = 1 + 2 \sum_{n=1}^{\infty} e^{-n^2 t},$$

and examine its behavior in the immediate neighborhood of the line of convergence, $\mathrm{Re}(t) = 0$. Set $t = \epsilon + \pi i p/q$, where ϵ is a small positive quantity and p and q are relatively prime positive integers. Then

$$f(\epsilon + \pi i p/q) = 1 + 2 \sum_{n=1}^{\infty} e^{-n^2 \epsilon} e^{-\pi i n^2 p/q}$$

(3)

$$= 1 + 2 \sum_{r=1}^{q} e^{-\pi i r^2 p/q} \left[\sum_{s=0}^{\infty} e^{-(r+sq)^2 \epsilon} \right]$$

upon taking account of the periodicity of $e^{-\pi i n^2 p/q}$ as a function of q.
    The function of ϵ

(4)
$$\sum_{s=0}^{\infty} e^{-(r+sq)^2 \epsilon}$$

behaves like the integral

(5)
$$\int_{0}^{\infty} e^{-(r+sq)^2 \epsilon} = \int_{r}^{\infty} e^{-w^2 \epsilon} \frac{dw}{q} \sim \frac{1}{q\sqrt{\epsilon}} \int_{0}^{\infty} e^{-w^2} dw = \frac{\sqrt{\pi}}{2q\sqrt{\epsilon}}$$

as $\epsilon \to 0$.
    Hence, asymptotically, as $\epsilon \to 0$, we have the equality

(6)
$$f\left(\epsilon + \frac{\pi i p}{q}\right) \sim \frac{\sqrt{\pi}}{q\sqrt{\epsilon}} S(p, q).$$

The periodicity of $e^{-\pi i n^2 p/q}$ as a function of q enables us to write

(7)
$$S(p, q) = \sum_{r=0}^{q-1} = \sum_{r=1}^{q}.$$

We now employ the transformation formula

(8)
$$f(t) = \left(\frac{\pi}{t}\right)^{1/2} f\left(\frac{\pi^2}{t}\right)$$

and repeat the process with the function $f(\pi^2/t)$.
    We have, for small ϵ,

(9)
$$\frac{\pi^2}{t} = \frac{\pi^2}{\epsilon + \pi i p/q} = \frac{\pi^2(\epsilon - \pi i p/q)}{\epsilon^2 + \pi^2 p^2/q^2} = \frac{\epsilon q^2}{p^2} - \frac{\pi i q}{p} + 0(\epsilon^2).$$

Hence, as above, as $\epsilon \to 0$,

$$(10) \qquad f\left(\frac{\pi^2}{t}\right) \sim \frac{\sqrt{\pi}}{q\sqrt{\epsilon}} S(-q, p).$$

Observe that the effect of the transformation formula of the theta function has been to invert the roles of p and q.

Taking account of the fact that

$$(11) \qquad \left[\frac{\pi}{(\epsilon + \pi i p/q)}\right]^{1/2} \sim e^{-\pi i/4}\left(\frac{q}{p}\right)^{1/2}$$

as $\epsilon \to 0$, we see that as $\epsilon \to 0$,

$$(12) \qquad \sqrt{\frac{\pi}{t}} f\left(\frac{\pi^2}{t}\right) \sim e^{-\pi i/4}\left(\frac{q}{p}\right)^{1/2}\left(\frac{\sqrt{\pi}}{q\sqrt{\epsilon}}\right) S(-q, p).$$

Hence, equating the two asymptotic results,

$$(13) \qquad \frac{1}{\sqrt{q}} \sum_{r=0}^{q-1} e^{-\pi i r^2 p/q} = \frac{e^{-\pi i/4}}{\sqrt{p}} \sum_{r=0}^{p-1} e^{\pi i r^2 q/p},$$

a remarkable functional equation.

Exercise

1. Derive from the formula above the law of quadratic reciprocity.

13. Eisenstein Series

Consider the function defined in the upper plane.

$$(1) \qquad f(z) = \sum{}' \frac{1}{(mz + n)^k},$$

where the prime indicates that m and n are not both zero simultaneously. A series of this type is called an Eisenstein series. It is clear that $f(z)$ satisfies the functional equation

$$(2) \qquad f(z) = z^k f(1/z).$$

More generally, we see that the function possesses a simple transformation rule under the transformation $(az + b)/(cz + d)$.

Functions of this form arise naturally from the Poisson summation formula or the Lipshitz identity. They also can be used to define general automorphic functions as was done by Poincaré.

14. Functional Equations

In the sections above, we put in the parameter additively. Now, let us put in the parameter in a multiplicative way. Consider the function defined by

$$(1) \qquad\qquad \cdot F(x) = \sum_{n=1}^{\infty} f(nx).$$

If we take the Mellin transform of both sides, we have

$$(2) \qquad\qquad M(F) = \zeta(s)M(f).$$

If we now invert, shift the contour, and use the functional equation for $\zeta(s)$ we obtain the Poisson summation formula. Naturally, the details require a bit of care. What is interesting about this method is that it shows that associated with every functional equation for a Dirichlet series is a summation formula.

15. Use of the Laplace Transform

We want a formula for the sum

$$(1) \qquad\qquad F(n) = \sum_{k=1}^{n} f(k).$$

We see that this function satisfies the equation

$$(2) \qquad\qquad F(n + 1) - F(n) = f(n + 1).$$

Hence, we are led to consider the difference equation

$$(3) \qquad\qquad F(x + 1) - F(x) = f(x), \quad x > 0.$$

This difference equation can clearly be treated by use of the Laplace transform. We have

$$(4) \qquad\qquad L(F) = \frac{L(f)}{e^s - 1}.$$

If we employ the Lipshitz identity, and the inversion formula, we find again the Poisson summation formula.

If we use the expansion of

(5)
$$\frac{1}{e^s - 1} - \frac{1}{s}$$

as a power series in s, we find a formula due to Euler-Maclaurin-Abel-Plana.

Exercises

1. Use a similar method for the vector-matrix equation $f(x + 1) = Af(x) + g(x)$.

2. Use the same method for the equation $u'(x + 1) = u(x) + f(x)$.

3. Use the Mellin transform for the q-difference equation $u(qx) = u(x) + f(x)$.

4. Use the Lipshitz identity to find the nth derivative of $1/(e^x - 1)$.

5. From this, obtain an estimate for the remainder term in the Taylor expansion of $1/(e^x - 1)$.

6. From this, obtain an estimate for the remainder in the Euler-Maclaurin sum formula.

Bibliography and Comments

Section 1

The general result was discussed by Poisson in 1827.
For a historical survey of summation formulae, which go back to Euler, Plana and Abel, see the monograph

Lindelof, E., *Le Calcul des Residus et Ses Applications,* Chelsea, New York, 1947.

A number of very elegant applications will be found in this treatise.

Section 3

The method we are following here is due to Bochner. See his book

Bocher, S., *Fourier Integral* (in German) New York, Chelsea, 1948.

The principle of Hecke is actually a particular case of a more embracing dictum to the effect that invariance under a group of transformations should always be made explicit, usually by means of an expansion in terms of group characters.

Good, I. J., *Analogues of Poisson's Summation Formula,* Amer. Math. Monthly **69** (4) (1962).

Let us note the following simple extension of (4) due to G. N. Watson:

$$\sum_{n=-\infty}^{\infty} g(n + s)e^{\pi i t(2n+s)} = \sum_{n=-\infty}^{\infty} f(n + t)e^{\pi i s(2n+i)},$$

where

$$g(x) = \int_{-\infty}^{\infty} f(y)e^{-2\pi ixy}dy.$$

Scott, S. A., *Some Applications of the Generalized Poisson-Jensen Formula,* Proc. Edinburgh Math. Soc. **6** (2) (1940), 151–156.

Section 4

An enormous amount of effort has been devoted to the study of conditions under which the Poisson summation formula is valid. The simple conditions above, although exceedingly restrictive, are sufficient for our present purposes.
The reader interested in more precise results may refer to

Titchmarsh, E. C., *Theory of Functions,* Oxford University Press, Oxford, 1939.

Mordell, L. J., *Poisson's summation formula and the Riemann zeta-Function,* J. Lond. Math. Soc. **4** (1928), 285-291.

Linfoot, E. H., *A Sufficiency Condition for Poisson's Formula,* J. London Math. Soc. **4** (1928), 54,

and the book by Bochner cited above.

Section 5

This proof of the transformation formula is due to Hamburger. See

Hamburger, H., *Über einige Beziehungen die mit der Funktionalgleichung der Riemannsche ζ-Funktion aequivalent sind,* Math. Ann. **85** (1922), 129.

The formula

$$\sum_{n=-\infty}^{\infty} e^{-\pi i n^2} = t^{-\frac{1}{2}} \sum_{n=-\infty}^{\infty} e^{-\pi n^2/t},$$

$$\sum_{n=-\infty}^{\infty} \exp\left[-\pi t \sum_{m=1}^{k} n_m^2\right] = t^{-k/2} \sum_{n=-\infty}^{\infty} \exp\left[-\frac{\pi}{t} \sum_{m=1}^{k} n_m^2\right].$$

This relation was extended by Hecke and Schoenberg to the identity

$$\sum_{n} P(n_1, n_2, \cdots, n_k) \exp\left(-\pi t \sum_{m=1}^{k} n_m^2\right)$$

$$= i^g t^{-k/2-g} \sum_{g} P(n_1, n_2, \cdots, n_k) \exp\left(-\frac{\pi}{t} \sum_{m=1}^{k} n_m^2\right),$$

where $P(x_1, x_2, \cdots, x_k)$ is any homogeneous polynomial of degree g satisfying the Laplacian $\Delta P = 0$.
  See

Schoenberg, B., Math. Ann. **116** (1939), 511.

These results were extended, using transform techniques, by Bochner.

Bochner, S., *Theta Functions with Spherical Harmonics,* Proc. Nat. Acad. Sci. Washington, D.C., **37** (1951), 804.

Section 10

For the details of the circle method, and other applications, all of which require skillful and delicate analysis, see

Hardy, G. H., *Ramanujan,* Chap. 9, Cambridge University Press, Cambridge, Massachusetts, 1927.

Section 11

For the details of the partition problem, see the book

Hardy, G. H. and E. M. Wright, *Introduction to the Theory of Numbers,* Clarendon Press, Oxford, 1960.

See also

Fischer, W., *One Dedekind's Function* $\eta(\tau)$, Pacific J. Math. **1** (1951), 83–95.
Rademacher, H., *On the Transformation of* $\log \eta(\tau)$, J. Indian Math. Soc. (N.S.) **19** (1955), 25–30.
Gupta, H., *On the Coefficients of the Powers of Dedekind's Modular Form,* J. London Math. Soc. **39** (1964), 433–440.

Section 12

This result is due to Schaar. The result of Gauss,

$$\sum_{r=0}^{q-1} e^{2\pi i r^2/q} = \frac{(1 - i^q)}{(1 - i)} \sqrt{q},$$

valid for an odd integer q, is deduced by taking $p = 2$. The proof given above is due to Landsberg,

Landsberg, M., *Zur Theorie der Gauss'schen Summen und der linearen Transformationen der Thetafunktionen,* Crelle's J. Math. **111** (1893).

For further history and a different method of proof, see

Lindelof, E., *Le Calcul des Residus,* Chelsea, New York, 1947.

The simplest proof using contour integration is given by Mordell in his paper in Acta Math. **61**.
See also

Lerch, M., *Zur Theorie der Gauss'schen Summen,* Math. Ann. **57** (1903), 554.
Mordell, L. J., *On the Reciprocity Formula for the Gauss's Sums in the Quadratic Field,* Proc. London Math. Soc. **20** (1921/22), 289.

The Gauss sum is a particular trigonometric sum arising in the study of cyclotomic sums—sums of roots of unity occurring naturally in the problem of constructing a regular polygon of *n* sides by means of ruler and compass. It is interesting to see that this problem, extended to the lemniscate, led Gauss to his independent discovery of elliptic functions.

Section 13

See

Fueter, R., *Vorlesungen über die Singularen Moduln und due Komplexe Multiplikation der Elliptischen Funktionen, Vol.* 1, Leipzig, 1924.

The strong impetus to the study of automorphic functions actually came from another direction as a result of studies by Fuchs on the ratio of solutions of particular second-order linear differential equations. See

Schlesinger, L., *Handbuch der Theorie der Linearen Differentialgleichung,* Leipzig, 1897.

That the Eisenstein series must be related to the theta functions is clear from the definition of the Weierstrassian elliptic functions. The importance of the series resides in the fact that similar series can be formed in more general situations involving functions of several complex variables where the corresponding elliptic or hyperelliptic functions cannot be defined in the same way. See

Hecke, E., *Mathematische Werke,* Göttingen, 1959.

where a number of papers containing discussion of generalized Eisenstein series may be found.

Section 14

A rigorous discussion is contained in

Ferrar, W. L., *Summation Formulae and their Relation to Dirichlet's Series* II, Compositio Math. **4** (1936/37), 394.

Ferrar's paper also contains a treatment of what modifications are necessary when $f(x)$ possesses various singularities.

For some interesting applications of the Poisson summation formula, see

Mordell, L. J., *Some Applications of Fourier Series in the Analytic Theory of Numbers,* Proc. Cambridge Philos. Soc. **34** (1928).

——, *Poisson's Summation Formula and the Riemann Zeta Function,* J. London Math. Soc. **4** (1929).

Guinand, A. P., *Some Finite Identities Connected with Poisson's Summation Formula,* Proc. Edinburgh Math. Soc. **12** (2) (1960/61), 17–25.

Ferrar, W. L., *Summation Formulae and Their Relations to Dirichlet Series*, Compositio Math. **1** (1935), 344.

The most famous formula of this type was used by Voronoi (Woronow) in his treatment of the sum $\Sigma_{n \leqslant N} d(n)$. Here $d(n)$, the Dirichlet divisor function, is the coefficient of n^{-s} in $\zeta^2(S)$.

$$\zeta^2(S) = \sum_{n=1}^{\infty} d(n)n^{-s}.$$

See

Voronoi, C., *Sur une Fonction Transcendante et ses Applications à la Summation de Quelques Séries*, Ann. Sci. École Norm. Sup. (Paris) **21** (3) (1904), 207-267, 459.

For some interesting series connected with the Gauss circle function $r(n)$ and the transformation formula

$$t^{\frac{1}{2}} \sum_{n=0}^{\infty} r(n)e^{-n\pi t} = t^{-\frac{1}{2}} \sum_{n=0}^{\infty} r(n)e^{-n\pi/t},$$

see

Kochliakov, S., Messeng. Math. **59** (1929), 1.

Further references will be found in the papers of Ferrar. A summary of recent results is given in

Sklar, A., *On Some Exact Formulae in Analytic Number Theory*, Report of the Institute of the Theory of Numbers, Boulder, Colorado, University of Colorado, June 21 - July 17, 1959.

Detailed analysis is necessary for the sums that arise from these summation formulas. See

Atkinson, F. V., *The Mean Value of the Zeta-Function on the Critical Line*, Proc. London Math. Soc. **47** (2) (1942), 174-200.
Dixon, A. L. and W. L. Ferrar, *Lattice-Point Summation Formulae*, Quart. J. Math. (Oxford Series) **2** (1931), 31-54.
Ferrar, W. L., *Summation Formulae and Their Relation to Dirichlet's Series*, Compositio Math. **1** (1935), 344-360.
——, *Summation Formulae and Their Relation to Dirichlet's Series*. II, Compositio Math. **4** (1937), 394-405.

Wilton, J. R., *An Approximate Functional Equation for the Product of Two ζ-Functions,* Proc. London Math. Soc. **31** (2) (1930), 11–17.

Bellman, R., *An Analog of an Identity Due to Wilton,* Duke Math. J. **16** (1949), 539–545.

See also,

Pearson, T. L., *Note on the Hardy-Landau Summation Formula,* Canad. Math. Bull. **8** (1965), 717–720.

Soni, K., *Some Relations Associated with an Extension of Koshliakov's Formula,* Proc. Amer. Math. Soc. **17** (1966), 543–551.

Nasim, C., *A Summation Formula Involving* $\sigma_k(n)$, $k > 1$, Canad. J. Math. **21** (1969), 951–964.

Berndt, B. C., *Arithmetical Identities and Hecke's Functional Equation,* Proc. Edinburgh Math. Soc. **16** (2) (1968/69), 221–226.

Nasim, C., *On the Summation Formula of Voronoi,* Trans. Amer. Math. Soc. **163** (1971), 35–45.

Section 15

See

Chowla, S., *On Some Formulae Resembling the Euler-Maclaurin Sum Formula,* Norske Vid. Selsk. Forh. (Trondheim) **34** (1961), 107–109.

For the Laplace transform see the book

Bellman, R. and K. L. Cooke, *Differential-Difference Equations,* Academic Press, New York, 1963.

7. Functional Equations

1. Introduction

In this chapter, we want to discuss the functional equations which determine the theta function. We have already seen how functional equations arise in the study of the gamma function and the zeta function. In the bibliography and comments at the end of the chapter, we will briefly discuss how functional equations arise in other parts of number theory. We shall give reference to work on differential difference equations.

2. Functional Equations Satisfied by Theta Functions

We begin with the function

$$\text{(1)} \qquad \theta\,(z,\,t) = \sum_{n=-\infty}^{\infty} e^{-n^2 \pi t} e^{2n\pi i z}.$$

This function is obviously periodic of period one. This means that it satisfies the equation

$$\text{(2)} \qquad \theta(z + 1,\, t) = \theta(z,\, t).$$

To obtain a further equation, we use the obvious fact

$$\text{(3)} \qquad \sum_{n} = \sum_{n+1}.$$

Richard Bellman, Analytic Number Theory: An Introduction   ISBN 0-8053-0452-5

If we simplify the right-hand side, we obtain the functional equation

(4) $$\theta(z + t, t) = e^{2\pi iz} e^{-\pi t} \theta(z, t).$$

Exercises

1. Find the functional equation for θ'/θ.

2. By using Rouche's theorem, find the number of zeros of θ in a fundamental parallelogram.

3. By evaluating $\int(\theta'/\theta)e^{2\pi niz}dz$ around the fundamental parallelogram find the Fourier series for θ'/θ. (Hermite)

4. Find analogous functional equations for

$$\sum_{-\infty < m,n < \infty} e^{-(am^2 + 2bmn + cn^2)} e^{2\pi miz + 2\pi niw},$$

where the quadratic form is positive definite.

5. Let $J(f, g)$ be the Jacobian of the two functions f and g. Find the functional equations satisfied by $J(f, g)/fg$ where f and g satisfy functional equations of the foregoing type.

6. Find the Fourier expansion for $J(f, g)/fg$ by considering the integral

$$\int (J(f, g)/fg)e^{-(2mi\pi z + 2ni\pi w)}\, dz\, dw,$$

where the integration is over the fundamental parallelepiped.

7. Show that the sum

$$\sum' \frac{1}{(m + in)^{2k}},$$

where the sum is over all values of m and n and the prime indicates that m and n are not simultaneously zero, can be evaluated using the theta function.

8. More generally, evaluate the sum

$$\sum' \frac{1}{(m + nt)^{2k}}.$$

(These sums are more easily evaluated using the Weierstrassian elliptic function.)

3. Determination of $\theta(z, t)$

What we now wish to show, conversely, is that these functional equations determine $\theta(z, t)$.

If we use the periodicity, we know that the function must have the form

$$(1) \qquad \sum_{n=-\infty}^{\infty} u_n(t)e^{2n\pi iz}.$$

If we use the second functional relation, we obtain a recurrence relation for $u_n(t)$. In this way we see that any solution of the two functional relations must be a multiple of $\theta(z, t)$. The multiplying factor, however, may depend upon t.

4. The Modular Transformation

Consider the function $\theta(z/t, 1/t)$. To obtain the required transformation, we write

$$(1) \qquad e^{\dfrac{-2\pi iz^2}{t}} \theta(z/t, 1/t) = f(z, t).$$

We see that the effect is to interchange the transformations $z + 1$ and $z + t$. Using the original functional equations, we see that $f(z, t)$ satisfies the equations

$$(2) \qquad \begin{aligned} f(z + 1, t) &= f(z, t), \\ f(z + t, t) &= e^{-\pi t}e^{2\pi iz}f(z, t). \end{aligned}$$

It follows that f is a multiple of θ.

Exercise

1. Write the zero of the theta function $z = h(t)$. Show that $h(t)$ satisfies the equation $h(t) = \text{th}(1/t)$.

5. The Determination of the Multiplier

It remains to determine the multiplier. As we shall see there are several ways of doing this.

One way is to choose a particular value of z. For example, we can set $z = 0$. Thus, we reduce the general case to the special case.

6. Partial Differential Equations

'Another approach which is much more powerful in many cases, is to use the theory of partial differential equations. In particular, we use the theory of parabolic differential equations, the heat equation or the diffusion equation. The fundamental observation is that

$$e^{-n^2 \pi t} e^{2n\pi i z}$$

is a solution of the partial differential equation.

(1)
$$\frac{\partial u}{\partial t} = \frac{\partial^2 u}{\partial z^2}.$$

Since the equation is linear, a sum of two solutions is also a solution. If we use this fact on the left and on the right, we obtain the modular transformation. One advantage of this approach is that it can be used in many other cases.

7. Generalizations

As we have indicated in the exercises, the method can be extended to many other cases. Thus, we have a systematic way of handling theta functions of several variables.

Also, the method can be extended to groups. We thus have a way of defining theta functions over groups. We shall not enter into this because it will take us too far from analytic number theory.

8. k-Dimensional Linear Spaces

Consider the function $f(z, t)$ which satisfies the functional equations

(1)
$$f(z + 1, t) = f(z, t),$$
$$f(z + t, t) = a(t)e^{2k\pi i z} f(z, t).$$

The same methods that were used above show that any function that satisfies these relations is a linear combination of k particular solutions.

If we consider the original functional equation, we see that the kth power satisfies relations of the above form. We see then the reason why there must exist relations between particular solutions.

Exercises

1. Obtain the functional equations for the square of θ (z, t).

2. From this show that there must be a linear relation between any three solutions.

3. How do we obtain the coefficients?

9. The Case $k = 2$

Consider the four functions determined by taking all the combinations of plus or minus in the functional equations

(1)
$$u(z + 1, t) = \pm u(z, t),$$
$$u(z + t, t) = \pm e^{\pi i t} e^{2\pi i z} u(z, t).$$

Let us prosaically call these functions f_1, f_2, f_3, and f_4. It will be recognized that f_1 is the function $\theta(z, t)$.

It will be seen that the squares of these functions satisfy the equations

(2)
$$f(z + 1, t) = f(z, t),$$
$$f(z + t, t) = e^{-2\pi i t} e^{4\pi i z} u(z, t).$$

Hence, there is a linear relation between any three squares. We can go further than this. Consider the functions $f_i(z + w, t)f_j(z - w, t)$, where i and j runs through the numbers 1, 2, 3, and 4 independently. We see that these functions satisfy the equations *in* (2). Consequently, we have that this function is a linear combination of any two squares. The coefficients depend upon w. However, we see that this function is symmetric in z and w. Hence, we have

(3) $\quad f_i(z + w, t)f_j(z - w, t) = af_k^2(z, t)f_k^2(w, t) + bf_l^2(z, t)f_l^2(w, t).$

The coefficients a and b depend upon $i, j, k,$ and l.

Exercises

1. Consider the 16 functions determined by all combinations of plus or minus in the functional equations

$$f(z + 1, w, t_1, t_2, t_3) = \pm f(z, w, t_1, t_2, t_3), \quad \text{etc.}$$

Prove that a linear relation exists between the squares of any five of these 16 functions.

2. More generally, consider the functions which satisfy

$$g(z + 1, t) = \omega^r g(z, t),$$

$$g(z + t, t) = \omega^s e^{\pi i t} e^{2\pi i z} g(z, t),$$

where ω is a root of unity.

3. By taking $z = w$, obtain a duplication formula.

4. In general, obtain a formula for multiplication of the argument by n, where n is an integer.

5. Obtain multi-dimensional analog.

10. Complex Multiplication

Consider the function $\theta(zt, t)$. We see that this function has a very simple transformation under the transformation $z + 1$. Suppose we consider the transformation $z + t$ and choose t to satisfy

(1) $$t^2 = at + b,$$

where the roots are complex, and a and b are integers. We see then that the function $\theta(zt, t)$ satisfies very simple transformations if t is chosen this way. This phenomenon is called complex multiplication. It was discovered by Abel and plays an important role in the theory of elliptic functions, algebraic number theory, and modular functions.

Exercise

1. Find analogs of complex multiplication for the higher-dimensional case.

11. Doubly-Periodic Functions

Consider the function defined by the quotient $f_i(z, t)/f_j(z, t)$. Using the functional transformations, we see that this function is doubly-periodic.

Thus, the theory of doubly-periodic functions may be made to depend upon the theta functions, as was done by Jacobi. One advantage of using the theta function is its rapid convergence. For theoretical purposes, in many ways the functions introduced by Weierstrass are superior. As usual, it is a good idea to have them both.

Miscellaneous Exercises

1. What is the general solution of

$$g(qa) = e^{\dfrac{2\pi i r}{p}} g(q)?$$

2. Consider the infinite product $f(q) = \Pi_n(1 + qx^n)$. Show that $f(q)$ satisfies the functional equation $f(qx) = (1 + x)f(q)$, and thus determine the power series expansion.

3. Consider in a like fashion the infinite product $f(q) = \Pi_n(1 - qx^n)$.

4. Obtain a functional equation for the infinite continued fraction

$$u(q) = \cfrac{x}{1 + qx \cfrac{}{\overline{1 + q^2 x \cfrac{}{\overline{} \\ \cdots}}}}$$

5. Show that this nonlinear q-difference equation can be converted into two linear q-difference equations by means of the transformation $u(q) = v(q)/w(q)$.

6. Let $f(m, n)$ be the number of steps in the Euclidean algorithm where $m > n > 1$. Prove that $f(m, n)$ satisfies

$$f(m, n) = 1 + f(n, m - [m/n]n).$$

See,
J. D. Dixon, *A Simple Estimate for the Number of Steps in the Euclidean Algorithm,* Amer. Math. Monthly **78** (1971), 374–376.

7. Consider the Newton interpolation formula

$$f(n) = a_0 + a_1 n + a_2 \frac{n(n-1)}{2} + \cdots.$$

Show that the coefficients may be determined by the operation

$$f(n + 1) - f(n) = \Delta f(n).$$

8. Show that a continuous limit of this formula is the Taylor series.

9. Consider the analog of the foregoing interpolation series

$$f(n) = a_0 + a_1 \frac{q^n - 1}{(q - 1)} + \cdots.$$

Show that the coefficients may be determined by the operation

$$\frac{f(n + 1) - f(n)}{q^n} = \delta f(n).$$

10. Use the foregoing interpolation formula on the function $\Pi_{n=1}^{n=N}(1 - q^k)$. (See,

Bellman, R., *A* q-*Version of the Newton Interpolation Formula and Some Eulerian Identities*, Boll. Un. Mat. Ital. **16** 1961, 285–287.)

11. Consider the function $f(x) = \Sigma_{n=0}^{\infty} x^{2^n}$. Prove that this function satisfies the equation $f(x) = x + f(x^2)$.

12. Prove that this functional equation determines the power series.

13. Consider the sequence

$$a_{n+1} = \frac{a_n + b_n}{2},$$

$$b_{n+1} = \sqrt{a_n b_n},$$

$$a_0 = 1, \quad b_0 = x, \quad a < x < \infty.$$

Prove that we have $a_n \geqslant b_n$. Show that the sequence a_n is monotone decreasing, and the second sequence b_n is monotone increasing. Hence, show that they both converge, and that they converge to a common limit $\phi(x)$, the arithmetical geometric mean of Gauss.

(This mean plays an important role in the evaluation of elliptic integrals, the Landen transformation. See

Whittaker, E. T. and G. M. Watson, *A Course of Modern Analysis*, 4th ed., Cambridge University Press, Cambridge, 1935.)

14. Show that $\phi(x)$ satisfies a modular transformation.
(This function can be made the basis of a theory of elliptic functions. See

Gauss, C. F., *Gesammelte Werke*.

15. Prove that

$$a_{n+1} - b_{n+1} = (\sqrt{a_n} - \sqrt{b_n})^2.$$

Hence, show that we have quadratic convergence.

Bibliography and Comments

Section 1

For a detailed discussion of the theory of theta functions and the connection with elliptic functions. See,

E. T. Whittaker and G. N. Watson, *A Course of Modern Analysis,* 4th ed., Cambridge University Press, Cambridge, 1935.

R. Bellman, *A Brief Introduction to Theta Functions,* Holt, Rinehart, and Winston, New York, 1961.

Many further references will be found therein.

Were we to start the theory of elliptic functions from scratch it would be most sensible, as pointed out to me by L. J. Mordell, to use the notation of Weber, *Lehrbuch der Algebra, Vol.* 3. The functions $\theta_{gh}(z, w)$ are introduced, possessing the functional equations

$$\theta_{gh}(z + w, w) = (-1)^h e^{-2\pi i(2z + w)}\theta_{gh}(z, w),$$

$$\theta_{gh}(z + 1, w) = (-1)^g \theta_{gh}(z, w).$$

The great advantage of this notation lies in the fact that the subscripts determine the functional equations, and make it quite easy to remember many important properties of the theta functions which are not at all evident in our notation. See,

Watson, G. N., *The Final Problem: An Account of the Mock Theta-functions,* J. London Math. Soc. **11** (1936), 55.

Dragonette, L. A., *Some Asymptotic Formulae for the Mock Theta Series of Ramanujan,* Trans. Amer. Math. Soc. **72** (1952), 474.

For functions related in form, but of less significance in the general theory—the "fake" theta functions—see

Rogers, L. J., Proc. London Math. Soc. (II), **16** (1917), 315.

and for another class of related functions, see

Basoco, M. A., *On Certain Arithmetical Functions Due to G. Humbert,* J. Math. Pures Appl. **9** (1947), p. 237.

The method of functional equations enters into many parts of analytic number theory. In particular, differential difference equations occur.

Differential difference equations play a prominent role in number theory, both in the study of the divisibility property of integers and in a heuristic proof of the prime number theorem due to Cherwell.
See,

N. G. DeBruijn, *On the Number of Positive Integers* < x *and Free of Prime Factors* > y, Nederl. Akad. Wetensch. Proc. Ser. A. **54** (1) (1951), 1–12.

V. Ramaswami, *On the Number of Integers* ⩽ X *and Free of Prime Divisors* > X^c, Science and Culture **13** (1948), 503.

V. Ramaswami, *On the Number of Integers* ⩽ X *and Free of Prime Divisors* > X^c, *and a problem of S. S. Pillai,* Science and Culture **13** (1948), 503.

V. Ramaswami, *On the Number of Positive Integers* < X *and Free of Prime Divisors* > X^c, Science and Culture **13** (1948), 465.

E. M. Wright, *The Linear Difference-Differential Equations with Constant Coefficients,* Proc. Roy. Soc. Edinburgh Sect. A **62** (1949), 387–393.

R. Bellman and B. Kotkin, *On the Numerical Solution of a Differential Difference Equation Arising in Analytic Number Theory,* Math. Comput. **16** (1962), 473–475.

S. Chowla and W. E. Briggs, *On the Number of Positive Integers* ⩽ x *all of Whose Prime Factors are* ⩽ y, Proc. Amer. Math. Soc. **6** (1955), 558–562.

The subject of numerical solution of differential difference equations is quite interesting and relatively unexplored. In particular, it is very interesting to see if we possess methods that use a small amount of computer storage. See,

R. Bellman, *Methods of Nonlinear Analysis, Vol.* I, Academic Press, New York, 1970.

The procedure given here follows,

R. Bellman, *Functional Equations and Theta-Functions.* I, Proc. Nat. Acad. Sci. **45** (1959), 853–854.

Sometimes we derive identities from modular functions. Sometimes we derive identities for modular functions from results in the theory of numbers. For the method of arithmetical paraphase, see

E. T. Bell, *Algebraic Arithmetic,* Amer. Math. Soc., New York, 1927.

Let a and b be two positive numbers, and let two sequences $\{a_n\}$ and $\{b_n\}$ be determined by the recurrence relations

$$a_{n+1} = \frac{a_n + b_n}{2}, \qquad b_{n+1} = \sqrt{a_n b_n}, \qquad n \geqslant 0,$$

where $a_0 = a_0$, $b_0 = b$. It is easy to show that a_n and b_n converge to a common limit, $M(a, b)$, which is known as the *arithmetic-geometric mean* of Gauss.

The importance of this transformation is twofold. In the first place, Gauss showed by means of a Landen transformation that

$$\int_0^{\pi/2} \frac{d\theta}{(a_1^2 \cos^2\theta + b_1^2 \sin^2\theta)^{1/2}} = \int_0^{\pi/2} \frac{d\theta}{(a^2 \cos^2\theta + b^2 \sin^2\theta)^{1/2}},$$

a result which yields, upon repeated application, the relation

$$\int_0^{\pi/2} \frac{d\theta}{(a^2\cos^2\theta + b^2\sin^2\theta)^{1/2}} = \frac{\pi/2}{M(a,b)}.$$

Secondly, Gauss showed that the theory of elliptic functions can be founded on the function $M(a, b)$.

For a discussion of these matters, see

Watson, G. N., *The Marquis and the Land Agent, a Tale of the Eighteenth Century,* Math. Gaz., **XVII** (1933), 5.
Gauss, C. F., *Werke,* vol. III, p. 352.
Whittaker, E. T. and Watson, G. N., *A Course of Modern Analysis,* Chap. XII, p. 533, Exer. 46, Cambridge University Press, Cambridge; 1935.

There are generalizations of this identity, valid for hyperelliptic integrals, but none of the same elegance.

Section 2

The number of zeros common to several functions was expressed by a real integral by Kronecker. This result was used by Birkhoff and Kellogg in their paper on fixed points in function space. See

R. Bellman, *Modern Mathematical Classics. I: Analysis,* Dover, New York, 1961.

See also,

L. Carlitz, *Some q-Identities Related to the Theta Functions,* Boll. Un. Mat. Ital. **17** (3) (1962), 172–178.

Section 5

For the multi-dimensional case, see

R. Bellman and R. S. Lehman, *The Reciprocity Formula for Multidimensional Theta Functions,* Proc. Amer. Math. Soc. **12** (1961), 954–961.

Section 6

For a classical discussion of the problem of determining when

$$\lim_{t\to 0} u(x, t) = f(x)$$

see

L. Fejer, *Math. Ann.* **58**, 1904, 51.

For further discussion of these matters, see

G. Doetsch, *Theorie und Anwendung der Laplace-Transformation,* Dover, New York, 1943.

The two forms of particular solutions, $e^{-n^2\pi^2 t}\sin n\pi x$, and $e^{(-x^2/4t)}t^{\frac{1}{2}}$, are manifestations of a general principle that partial differential equations have similarity solutions for small times and separation-of-variables solutions for large t.

The functional equation for the theta functions, particular solutions of the heat equation, is a special case of a far more general transformation formula valid for general solutions of the heat equation.

Let $u(x, y, z, t)$ be a particular solution of the three-dimensional heat equation

$$\frac{1}{a^2}\frac{\partial u}{\partial t} = \frac{\partial^2 u}{\partial x^2} + \frac{\partial^2 u}{\partial y^2} + \frac{\partial^2 u}{\partial z^2}.$$

Then the function

$$U(x, y, z, t) = t^{-3/2}u\left(\frac{x}{t}, \frac{y}{t}, \frac{z}{t}, \frac{-1}{t}\right)\exp-\left(\frac{x^2 + y^2 + z^2}{4at}\right)$$

is also a solution. There are corresponding results for the n-dimensional heat equation, and analogous results for the wave equation.

We have presented the foregoing method of establishing transformation formulae, since it has been used by Maass and Siegel to establish some important identities which to date have not been derived by any other means. See

H. Maass, *Uber eine neue Art von Nichtanalytischen Automorphen Funktionen und die Bestimmung Dirichletschen Reihen durch Funktional-gleichungen,* Math. Ann. **121** (1949), 141.

C. L. Siegel, *Indefinite Quadratische Formen und Funktionentheorie.* I, Math. Ann. **124** (1951).

The heat equation is also the diffusion equation. This means that there is a connection with probability theory. This was utilized by Polya, see

G. Polya, *Elementarer Beweis einer Theta formel,* Sitzungsber. Phys. -Math. Klasse, Berlin, 1927, pp. 158–161.

Section 7

For the relevant theory of groups, see

L. S. Pontrjagin, *Topological Groups*, Princeton University Press, Princeton, New Jersey, 1966.

Section 10

Complex multiplication played an important part in the "Jugendtraum" of Kronecker.

Section 11

An excellent discussion of the doubly periodic function is in the book by E. T. Whittaker and G. M. Watson. See

E. T. Whittaker and G. M. Watson, *A Course of Modern Analysis*, 4th ed., Cambridge University Press, Cambridge, 1935.

For the theory of multiply periodic functions, see the book

G. F. Baker, *An Introduction to the Theory of Multiply Periodic Functions*, Cambridge University Press, Cambridge, 1907.

8. The Euler ϕ Function

1. Introduction

In this chapter, we will study one of the fundamental functions of number theory, the Euler φ function.

First, we will establish the fundamental multiplicative property. Using this, we obtain an Euler product from which we will derive a very simple representation. Using this representation, it is easy to derive various mean values.

Finally, at the end of the chapter, we shall discuss the Ramanujan function.

2. The Multiplicative Property

In this section, we shall derive the fundamental multiplicative property of the Euler φ function. We use a simple but useful technique which appears in many parts of number theory. We calculate our quantity by two different methods and equate the results.

It is convenient to calculate the number of integers, $\leqslant mn$, which have a factor in common with the product mn (where m and n are relatively prime).

On one hand, this number is

(1)
$$mn - \varphi(mn).$$

On the other hand, this number is

(2)
$$(m - \varphi(m))n + m(n - \varphi(n)) - (m - \varphi(m))\,(n - \varphi(n)).$$

The third term arises because the first two terms count several quantities

Richard Bellman, Analytic Number Theory: An Introduction   ISBN 0-8053-0452-5

twice. Equating the two expressions we obtain, after some simplification, the fundamental result

(3) $$\varphi(mn) = \varphi(m)\varphi(n), \quad (m, n) = 1.$$

It is clear that

(4) $$\varphi(p^a) = p^{a-1}(p - 1).$$

Hence, using the fundamental theorem of arithmetic, we have

(5) $$\varphi(n) = \prod_{p \mid n} p^{a-1}(p - 1).$$

Hence, we have

(6) $$\frac{\varphi(n)}{n} = \prod_{p \mid n} \left(1 - \frac{1}{p}\right).$$

Exercises

1. Show that $\overline{\lim}_{n \to \infty} \varphi(n)/n = 1$.

2. Show that $\underline{\lim}_{n \to \infty} \varphi(n)/n = 0$.

3. Show that if $(m, n) = 1$, $1 \leqslant m < n$, then $n - m$ is a number of the same form.

4. Using this fact, evaluate the sum $\Sigma_{(m,n)=1} m$ where $1 \leqslant m < n$.

5. Define $S_k = \Sigma_{(m,n)=1} m^k$, where $1 \leqslant m < n$. By considering the' sums S_3 and S_4 evaluate S_2 and S_3.

6. Generalize, to evaluate S_k.

7. Show that

$$\sum_{n=1}^{\infty} \frac{\varphi(n)x^n}{n} = \sum_{k=1}^{\infty} \frac{\mu(k)}{k} \frac{x^k}{(1 - x^k)}.$$

3. The Euler Product

Using the fundamental multiplicative property, we have the relation

$$\sum_{n=1}^{\infty} \frac{\varphi(n)}{n^s} = \prod_p \left(1 + \frac{\varphi(p)}{p^s} + \cdots\right)$$

$$= \prod_p \left(1 + \frac{p-1}{p^s} + \cdots\right)$$

(1)
$$= \prod_p \left(1 + \frac{(p-1)}{p^s} \Big/ (1 - 1/p^s)\right)$$

$$= \prod_p \left(\frac{1 - \frac{1}{p^{s+1}}}{1 - \frac{1}{p^s}}\right)$$

$$= \frac{\zeta(s-1)}{\zeta(s)} \text{ for } Re(s) > 2.$$

From this we derive the representation

(2)
$$\frac{\varphi(n)}{n} = \sum_{k|n} \frac{\mu(k)}{k}.$$

The quantity $\varphi(n)/n$ is so useful to consider that we shall call it $f(n)$.

Exercise

1. Show that the two representations of $\varphi(n)$ are equivalent.

4. The Mean Value of $\varphi(n)$

Using the representation above, we have

(1)
$$\sum_{n=1}^{N} f(n) = \sum_{n=1}^{N} \left(\sum_{k|n} \mu(k)/k\right).$$

Inverting the order of summation we have

(2)
$$\sum_{k=1}^{N} \frac{\mu(k)}{k} \left[\frac{N}{k}\right].$$

using the relation

(3) $0 \leqslant x - [x] < 1,$

we have

(4) $N \displaystyle\sum_{k=1}^{N} \frac{\mu(k)}{k^2} + 0\left(\sum_{k=1}^{N} \frac{1}{k}\right).$

This is a remarkably sharp estimate. Using the estimates we derived in Chap. 1, we can write this as

(5) $\dfrac{6N}{\pi^2} + 0(\log N).$

Exercises

1. Derive an estimate for $\Sigma_{k=1}^{N} \varphi(k)$.

2. Derive the constant represented by the symbol 0.

5. The Mean Value of $\varphi(n)^2$

To obtain the order of magnitude of the mean value of $f(n)^2$, we proceed as follows. We write,

(1) $\displaystyle\sum_{k=1}^{N} f(k)^2 = \sum_{k=1}^{N} f(k)\left(\sum_{r|k} \mu(r)/r\right).$

Inverting the summations, this may be written

(2) $\displaystyle\sum_{k=1}^{N} \frac{\mu(k)}{k}\left(\sum_{\substack{n \equiv 0(k) \\ 1 \leqslant k \leqslant N}} f(n)\right)$

Using the results of the previous section we see that the required mean value has the form

(3) $c_1 N + 0(\log N^2)$

where c_1 is a constant.

Exercises

1. Evaluate in this way the sum $\Sigma_{k=1}^{N} f(k)^3$.

2. Find the mean value of $\varphi(n)^2$.

3. Find the order of magnitude of the Ingham sum $\Sigma_{n=1}^{N} f(n)f(n+a)$.

4. Find the order of magnitude of the Ingham sum $\Sigma_{n=1}^{N} \varphi(n)\varphi(n+a)$.

6. An Alternate Method

It is clear that we may proceed inductively to find the order of magnitude of the sums

$$(1) \qquad \sum_{n=1}^{N} \varphi(n)^k.$$

However, the arithmetic soon gets rather unpleasant. Let us then give a different approach.

We have

$$
\sum_{n=1}^{\infty} \frac{f(n)^k}{n^s} = \prod_{p}\left(1 + \frac{f(p)^k}{p^s} + \cdots\right)
$$

$$
= \prod_{p}\left(1 + \frac{(1-\frac{1}{p})^k}{p^s} + \cdots\right)
$$

$$(2) \qquad = \prod_{p}\left(1 + \frac{(1-\frac{1}{p})^k/p^s}{1 - \frac{1}{p^s}}\right)$$

$$
= \zeta(s)g(s),
$$

where $g(s)$ has an absolutely convergent Dirichlet series for $s > \frac{1}{2}$.

Hence, we may write

$$(3) \qquad f(n)^k = \sum_{r\mid n} a_r,$$

where a_r is absolutely convergent. Thus we have

(4)
$$\sum_{n=1}^{N} f(n)^k = \sum_{r=1}^{N} a_r \left[\frac{N}{r} \right] \sim N \sum_{r=1}^{\infty} a_r/r.$$

We leave it as an exercise to evaluate the error term.

Exercise

1. Find the order of magnitude of $\sum_{n=1}^{N} \varphi(n)^k$.

7. The Mean Value of $\varphi(p(n))$

In what follows, we assume that $p(n)$ has no multiple factors.

Let us now consider the mean value of $\varphi(p(n))$ where $p(n)$ is a polynomial. We have

(1)
$$\sum_{n=1}^{N} f(p(n)) = \sum_{n=1}^{N} \left(\sum_{k \mid p(n)} \mu(k)/k \right).$$

Inverting the order of the summations, we have

$$\sum_{k=1}^{N} \frac{\mu(k)}{k} \sum_{\substack{f(n)\equiv 0(k) \\ 1 \leqslant n \leqslant N}} 1$$

(2)
$$= \sum_{k=1}^{N} \frac{\mu(k)\rho(k)}{k} \left[\frac{N}{k} \right] + 0(\log N)$$

$$= N \sum_{k=1}^{\infty} \frac{\mu(k)\rho(k)}{k^2} + 0(\log N).$$

8. The Ramanujan Function

Let us now consider an important function which may be considered a generalization of $\varphi(n)$.

We define

(1)
$$c_q(n) = \sum_{(m,q)=1} e^{\frac{2\pi i m n}{q}}.$$

This function has some interesting properties which we shall discuss further in Chap. 9.

Miscellaneous Exercises

1. Prove that $c_q(n)$ is a multiplicative function.

2. Evaluate $c_q(n)$ in terms of $\varphi(n)$. (Holder)

3. Prove that the $c_q(n)$ are orthogonal. (Carmichael)

Bibliography and Comments

Section 1

The Euler φ function is one of the most interesting functions of number theory. It has many extensions. See

Nageswara, R. K., *On Extensions of Euler's φ Function,* Math. Student **29** (1961), 121-126.

——, *A Note on the Extension of Euler's φ Function,* Math. Student **29** (1961), 33-35.

——, *On Jordan Function and Its Extension,* Math. Student **29** (1961), 25-28.

Alder, H. L., *A Generalization of the Euler φ-Function,* Amer. Math. Monthly **65** (1958), 690-692.

Satyanarayana, U. V. and K. Pattabhiramasastry, *A Note on the Generalized φ-Functions,* Math. Student **33** (1965), 81-83.

Shockley, J. E. and R. J. Hursey, *On the Totient Functions of Jordan and Zsigmondy,* Amer. Math. Monthly **73** (1966), 608-610.

Cohen, E., *Some Totient Functions,* Duke Math. J. **23** (1956), 515-522.

McCarthy, P. J., *Notes on Some Arithmetical Sums,* Boll. Un. Mat. Ital. **21** (3) (1966), 239-242.

Tanaka, K., *A Generalization of Euler's φ Function,* Res. Rep. Tokyo Electrical Engrg. College **15** (1967), pp. 24-29.

Cohen, E., *A Generalized Euler φ-Function,* Math. Mag. **41** (1968), 276-279.

Matsuda, R., *On a Generalization of Euler's φ-Function,* Bull. Fac. Sci. Ibaraki Univ. Ser. A., No. 2-2 (1970), pp. 19-21.

Stevens, H., *Generalization of the Euler φ-Function,* Duke Math. J. **38** (1971), 181-186.

Bell, E. T., *Interpolated Denumerants and Lambert Series,* Amer. J. Math. **65** (1943), 382-386.

Haviland, E. K., *An Analogue of Euler's φ-Function,* Duke Math. J. **11** (1944), 869-872.

Siez, C. -S., *A General Expression of Euler's φ-Function,* J. Indian Math. Soc. (N.S.) **8** (1944), 91-94.

Erdös, P., *Some Remarks on Euler's φ-Function and Some Related Problems,* Bull. Amer. Math. Soc. **51** (1945), 540-544.

Hanumanthachari, J., *Certain Generalizations of Nagell's Totient Function and Ramanujan's Sum,* Math. Student **38** (1970), 183-187.

Donnelly, H., *On a Problem Concerning Euler's Phi-Function,* Amer. Math. Monthly **80** (1973), 1029-1031.

Wall, C. R., *Density Bounds for Euler's Function,* Math. Comp. **26** (1972), 779-783.

Grosswald, E., *Contribution to the Theory of Euler's Function $\varphi(x)$,* Bull. Amer. Math. Soc **79** (1973), 337-341.

We have avoided the interesting question of the distribution of values of the φ-function. See

Erdös, P. and H. N. Shapiro, *The Existence of a Distribution Function for an Error Term Related to the Euler Function,* Canad. J. Math. 7 (1955), 63–75.

Lehmer, D. H., *The Distribution of Totatives,* Canad. J. Math. 7 (1955), 347–357.

McCarthy, P. J., *Note on the Distributions of the Totatives,* Amer. Math. Monthly **64** (1957), 585–586.

Erdös, P., *Some Remarks on a Paper of McCarthy,* Canad. Math. Bull. **1** (1958), 71–75.

Subbarao, M. V. and D. Suryanarayana, *On an Identity of Eckford Cohen,* Proc. Amer. Math. Soc. **33** (1972), 20–24.

Wintner, A., *Number-Theoretical Almost-Periodicities,* Amer. J. Math. **67** (1945), 173–193.

Klee, V. L., Jr., *On the Equation $\varphi(x) = 2m$,* Amer. Math.Monthly **46** (1946), 327–328.

Bateman, P. T., *The Distribution of Values of the Euler Function,* Acta Arith. **21** (1972), 329–345.

We have made no attempt to collect systematically all the interesting identities of number theory. Some of these may be found in the following papers where further references are given.

Carlitz, L., *An Arithmetic Sum Connected with the Greatest Integer Function,* Norske Vid. Selsk. Forh. (Trondheim) **32** (1959), 24–30.

——, *An Arithmetic Sum Connected with the Greatest Integer Function,* Math. Scand. **8** (1960), 59–64.

——, *An Arithmetic Sum Connected with the Greatest Integer Function,* Arch. Math. **12** (1961), 34–42.

——, *Some Finite Summation Formulas of Arithmetic Character,* Publ. Math. Debrecen **6** (1959), 262–268.

Cohen, E., *The Elementary Arithmetical Functions,* Scripta Math. **25** (1960), 221–227.

McCarthy, P. J., *The Generation of Arithmetical Identities,* J. Reine Angew. Math. **203** (1960), 55–63.

Carlitz, L., *Some Finite Summation Formulas of Arithmetic Character.* II, Acta Math. Acad. Sci. Hungar. **11** (1960), 15–22.

Cohen, E., *Arithmetical Functions Associated with the Unitary Divisors of an Integer,* Math. Z. **74** (1960), 66–80.

Apostol, T. M., *Identities Involving the Coefficients of Certain Dirichlet Series,* Duke Math. J. **18** (1951), 517–525.

Cohen, E., *Representations of Even Functions* (mod r). II. *Cauchy Products,* Duke Math. J. **26** (1959), 165–182.

Amitsur, S. A., *On Arithmetic Functions,* J. Analyze Math. **5** (1956/57), 273–314.

Cohen, E., *An Arithmetical Inversion Principle,* Bull. Amer. Math. Soc. **65** (1959), 335–336.

Morgado, J., *Unitary Analogue of the Nagell Totient Function,* Portugal. Math. **21** (1962), 221–232.

McCarthy, P. J., *Some More Remarks on Arithmetical Identities,* Portugal. Math. **21** (1962), 45–57.

——, *Busche-Ramanujan Identities,* Amer. Math. Monthly **67** (1960), 966–970.

Makowski, A. M., *On Some Equations Involving $\varphi(n)$ and $\sigma(n)$,* Amer. Math. Monthly **67** (1960), 668–670.

Shapiro, H., *An Arithmetic Function Arising from the φ Function,* Amer. Math. Monthly **50** (1943), 18–30.

Gunderson, N. G., *Some Theorems on the Euler φ-Function,* Bull. Amer. Math. Soc. **49** (1943), 278–280.

Cohen, E., *A Property of Dedekind's φ-Function,* Proc. Amer. Math. Soc. **12** (1961), 996.

Amitsur, S. A., *Some Results on Arithmetic Functions,* J. Math. Soc. Japan **11** (1959), 275–290.

Balasubramanian, N., *Some Identities in Number Theoretic Analysis,* Math. Student **29** (1961), 89–92.

Cohen, E., *Arithmetical Notes. VI, Simultaneous Binary Compositions Involving Coprimes Pairs of Integers,* Michigan Math. J. **9** (1962), 277–282.

——, *Series Representations of Certain Type of Arithmetical Functions,* Osaka Math. J. **13** (1961), 209–216.

——, *Arithmetical Functions of a Greatest Common Divisor. II, An Alternative Approach,* Boll. Un. Mat. Ital. **17** (3) (1962), 349–356.

Cohen, E., *Arithmetical Functions of a Greatest Common Divisor. III, Cesáro Divisor Problem,* Proc. Glasgow Math. Assoc. **5** (1961), 67–75.

——, *Arithmetical Notes, XIII, A Sequel to Note IV,* Elem. Math. **18** (1963), 8–11.

——, *Unitary Products of Arithmetical Functions,* Acta Arith. **7** (1961/62), 29–38.

Narkiewicz, W., *On a Summation Formula of E. Cohen,* Colloq. Math., **11** (1963), 85–86.

Cohen, E., *Arithmetical Notes. XII, A Sequel to Note VI,* Norske Vid. Selsk. Forh. (Trondheim) **36** (1963), 10–15.

Makowski, A., *Remark on the Euler Totient Function,* Math. Student **31** (1963), 13–14.

Cohen, E., *Arithmetical Notes. VII, Some Classes of Even Functions (mod r),* Collect. Math. **16** (1964), 81–87.

Duncan, R. L., *Generating Functions for a Class of Arithmetical Functions,* Amer. Math. Monthly **72** (1965), 882–884.

Nageswara, R. K., *On the Unitary Analogues of Certain Totients,* Montash. Math. **70** (1966), 149–154.

Carlitz, L., *Arithmetic Functions in an Unusual Setting*. II, Duke Math. J. **34** (1967), 757-759.

Rao, K. Nageswara, *Unitary Class Division of Integers* mod n *and Related Arithmetical Identities*, J. Indian Math. Soc. (N.S.) **30** (1966), 195-205.

Buschman, R. G., *Identities Involving Products of Number-Theoretic Functions*, Proc. Amer. Math. Soc. **25** (1970), 307-309.

Suryanarayana, D., *The Greatest Divisor of* n *which is prime to* k, Math. Student **37** (1969), 147-157.

Carlitz, L., *The Greatest Integer Function*, Delta (Waukesha) **1** (1968/69), 1-12.

Cohen, E., *On the Mean Parity of Arithmetical Functions*, Duke Math. J. **36** (1969), 659-668.

Wegner, K. W. and S. R. Savitzky, *Solutions of* $\varphi(x) = n$, *where* φ *is Euler's* φ-*Function*, Amer. Math. Monthly **77** (1970), 287.

Galambos, J., *Distribution of Arithmetical Functions. A Survey*, Ann. Inst. N. Poincaré Sect. B (N.S.) **6** (1970), 281-305.

Nageswara, R. K., *On Certain Arithmetical Sums, The Theory of Arithmetic Functions* (Proc. Conf., Western Michigan University, Kalamazoo, Michigan, 1971), pp. 181-192. Lecture Notes in Math., Vol. 251, Springer, Berlin, 1972.

Srinivasan, B. R., *On the Number of Abelian Groups of a Given Order*, Acta Arith. **23** (1973), 195-205.

Pomerance, C., *On Carmichael's Conjecture*, Proc. Amer. Math. Soc. **43** (1974), 297-298.

Jean-Marie De Koninck, *On a Class of Arithmetical Functions*, Duke Math. J. **39** (1972), 807-818.

Steinig, J., *On an Integral Connected with the Average Order of a Class of Arithmetic Functions*, J. Number Theory **4** (1972), 463-468.

Many interesting questions occur when we ask for iteration of $\varphi(n)$. See

Erdös, P., *Some Remarks on the Iterates of the* φ *and* σ *Functions*, Colloq. Math. **17** (1967), 195-202.

Parnami, J. C., *On Iterates of Euler's* φ-*Function*, Amer. Math. Monthly **74** (1967), 967-968.

Lal, Mohan, *Iterates of a Number-Theoretic Function*, Math. Comp. **23** (1969), 181-183.

Shapiro, H. N., *On the Iterates of a Certain Class of Arithmetic Functions*, Comm. Pure Appl. Math. **3** (1950), 259-272.

White, G. K., *Iterations of Genralized Euler Functions*, Pacific J. Math. **12** (1962), 777-783.

Gruzewski, A. and A. Schinzel, *Sur les Itérations d'une Fonction Arithmétique*, Prace Mat. **11** (1968), 279-282.

Kátai, I., *On the Iteration of the Divisor Function*, Publ. Math. Debrecen **16** (1969), 3–15.

Erdös, P. and M. V. Subbarao, *On the Iterates of Some Arithmetic Functions, The Theory of Arithmetic Functions* (Proc. Conf., Western Michigan University, Kalamazoo, Michigan, 1971), pp. 119–125. Lecture Notes in Math., Vol. 251, Springer, Berlin, 1972.

Lal, M., *Iterates of the Unitary Totient Function*, Math. Comp. **28** (1974), 301–302.

Mills, W. H., *Iteration of the φ Function*, Amer. Math. **50** (1943), 547–549.

Section 4

See

Segal, S. L., *A Note on Normal Order and the Euler φ-Function*, J. London Math. Soc. **39** (1964), 400–404.

Kesava, M. P., *On the Sum $\Sigma(a - 1, n)$, $[(a, n) = 1]$*, J. Indian Math. Soc. (N.S.) **29** (1965), 155–163.

de Bruijn, N. G. and J. H. van Lint, *On Partial Sums of $\Sigma_{d/m}\varphi(d)$*, Simon Stevin **39** (1965/66), 18–22.

Segal, S. L., *On Non-decreasing Normal Orders*, J. London Math. Soc. **40** (1965), 459–466.

Miech, R. J., *An Asymptotic Property of the Euler Function*, Pacific J. Math. **19** (1966), 95–107.

MacLeod, R. A., *The Minimum of $\Phi(x)/x^2$*, J. London Math. Soc. **42** (1967), 352–660.

Steinig, J., *The Changes of Sign of Certain Arithmetical Error-Terms*, Comment. Math. Helv. **44** (1969), 385–400.

Berndt, B. C., *On the Average Order of Ideal Functions and Other Arithmetical Functions*, Bull. Amer. Math. Soc. **76** (1970), 1270–1274.

Swetharanyam, S., *Asymptotic Expressions for Certain Type of Sums Involving the Arithmetic Functions in the Theory of Numbers*, Math. Student **28** (1960), 9–28.

Cohen, E., *An Elementary Method in the Asymptotic Theory of Numbers*, Duke Math. J. **28** (1961), 183–192.

Chandrasekharan, K. and R. Narasimhan, *An Average Order of Arithmetical Functions, and the Approximate Functional Equation for a Class of Zeta-Functions*, Rend. Mat. **21** (5) (1962), 354–363.

Cohen, E., *Arithmetical Notes. VIII, An Asymptotic Formula of Renyi*, Proc. Amer. Math. Soc. **13** (1962), 536–539.

——, *Averages of Completely Even Arithmetical Functions over Certain Types of Plane Regions*, Ann. Mat. Pura Appl. **59** (4) (1962), 165–177.

Grosswald, E., *Oscillation Theorems of Arithmetical Functions*, Trans. Amer. Math. Soc. **126** (1967), 1–28.

Dodson, M. M., *The Average Order of Two Arithmetical Functions,* Acta Arith. **16** (1969/70), 71–84.

Lehmer, D. H., *On the Compounding of Certain Means,* J. Math. Anal. Appl. **36** (1971), 183–200.

Proschan, J. H., *On the Changes of Sign of a Certain Class of Error Functions,* Acta Arith. **17** (1970/71), 407–430.

Berndt, B. C., *On the Average Order of a Class of Arithmetical Functions.* II, J. Number Theory 3 (1971), 288–305.

Rodabaugh, D. J., *A Maximization Problem Involving Totients, Nieuw Arch. Wisk.* **17** (3) (1969), 133–141.

Fogels, E., *On Average Values of Arithmetic Functions,* Proc. Cambridge Philos. Soc. **37** (1941), 358–372.

Annapurna, U., *Inequalities for* $\sigma(n)$ *and* $\varphi(n)$, Math. Mag. **45** (1972), 187–190.

Suryanarayana, D., *On* $\Delta(x, n) = \varphi(x, n) - x\varphi(n)/n$, Proc. Amer. Math. Soc. **44** (1974), 17–21.

Section 8

The function of Ramanujan can be extensively generalized. It has been extended by many authors. See

Cohen, Ekford, *An Extension of Ramanujan's Sum.* I, Duke Math. J. 16 (1949), 85–90.

——, *An Extension of Ramanujan's Sum.* II, *Additive Properties,* Duke Math. J. **22** (1955), 543–550.

——, *An Extension of Ramanujan's Sum.* III, *Connections with Totient Functions,* Duke Math. J. **23** (1956), 623–630.

Venkataraman, C. S., *Further Application of the Identical Equation to Ramanujan's Sum* $C_M(N)$ *and Kronecker's Function* p(M, N), J. Indian Math. Soc. (N.S.) **10** (1946), 57–61.

McCarthy, P. J., *Some Properties of the Extended Ramanujan Sums,* Arch. Math. **11** (1960), 253–258.

Sugunamma, M., *Eckford Cohen's Generalization of Ramanujan's Trigonometrical Sum* C(n, r), Duke Math. J. **27** (1965), 323–330.

Venkataraman, C. S., *On Von Sterneck-Ramanujan Function,* J. Indian Math. Soc. (N.S.) **13** (1949), 65–72.

Nicol, C. A., *Some Formulas Involving Ramanujan Sums,* Canad. J. Math. **14** (1962), 284–286.

Nageswara, R. K. and K. Rao, *Generalization of a Theorem of Eckford Cohen,* Math. Student **29** (1961), 83–87 (1962).

Horadam, E. M., *Ramanujan's Sum for Generalized Integers,* Duke Math. J. **31** (1964), 697–702.

Cohen, E., *An Identity Related to the Dedekind-von Sterneck Function,* Amer. Math. Monthly **69** (1962), 213–215.

Subba, M. V. Rao (M. V. Subbarao) and V. C. Harris, *A New Generalization of Ramanujan's Sum,* J. London Math. Soc. **41** (1966), 595–604.

Horadam, E. M., *Ramanujan's Sum and Its Application to the Enumerative Functions of Certain Sets of Elements of an Arithmetical Semigroup,* J. Math. Sci. **3** (1968), 47–70.

——, *Ramanujan's Sum and Nagell's Totient Function for Arithmetical Semi-Groups,* Math. Scand. **22** (1968), 269–281.

Lahiri, D. B., *Some Arithmetical Identities for Ramanujan's and Divisor Functions,* Bull. Austral. Math. Soc. **1** (1969), 307–314.

Rao, K. Nageswara, *Some Identities Involving an Extension of Ramanujan Sum,* Norske Vid. Selsk. Forh. (Trondheim) **40** (1967), 18–23.

Suryanarayana, D., *A Property of the Unitary Analogue of Ramanujan's Sum,* Elem. Math. **25** (1970), 114.

Wintner, A., *On a Statistics of the Ramanujan Sums,* Amer. J. Math. **64** (1942), 106–114.

Apostol, T. M., *Arithmetic Properties of Generalized Ramanujan Sums,* Pacific J. Math. **41** (1972), 281–293.

Venkataraman, C. S. and R. Sivaramakrishnan, *An Extension of Ramanujan's Sum,* Math. Student **40A** (1972), 211–216.

9. The Divisor Function

1. Introduction

In this chapter, we study the divisor function.

In the second section, we recall some facts about this function. In the third section, we consider its mean value. In the fourth section, using a simple geometric method, we show that the error term can be considerably improved. In the fifth section, we show an analytic equivalent of the geometric result which we use in a section below. In the sixth section, we consider the mean value of $d(n)^2$. In the next section, we show the connection with certain Diophantine equations. In the next section, we give the famous Perron sum-formula. Then we show how this formula may be extended by a summability technique. Then we turn to the mean value of $d(an^2 + bn + c)$. In the next section, we show why the method does not work for $d(p(n))$, where $p(n)$ is a general polynomial. In the final section we show how Ramanujan expansions can be used for the evaluation of the order of magnitude of $\sigma(p(n))$.

2. The Divisor Function

Let us now consider some facts about the divisor function. This function is the number of the divisors of n. Analytically, we may write

(1)
$$d(n) = \sum_{k \mid n} 1.$$

A generating function is

Richard Bellman, Analytic Number Theory: An Introduction   ISBN 0-8053-0452-5

(2)
$$\zeta^2(s) = \sum_{n=1}^{\infty} \frac{d(n)}{n^s} .$$

Since $d(n)$ is multiplicative, we have

(3)
$$d(n) = \prod_{p} (1 + \alpha_i),$$

where

(4)
$$n = \Pi p_i^{\alpha_i}.$$

In this chapter we shall not use the multiplicative result.

Exercises

1. Show that $d(n) = 0(n^\epsilon)$ for any $\epsilon > 0$.

2. Show that there are integers n such that $d(n) \geqslant (\log n)^k$ for any k greater than 0.

3. Find the order of magnitude of $\Sigma_{n=1}^{N} d(an)$.

4. Find the order of magnitude of $\Sigma_{n=1}^{N} d(an + b)$.

5. Find the order of magnitude of $\Sigma_{n=1}^{N} r(n)$.

6. Show that

$$\sum_{n=1}^{\infty} \frac{x^n}{n^s(1 - x^n)} = \sum_{n=1}^{\infty} \sigma_s(n)x^n.$$

7. Show that $\sigma_s(n) = n^s \sigma_{-s}(n)$.

3. The Mean Value of $d(n)$

We have

(1)
$$\sum_{n=1}^{N} d(n) = \sum_{n=1}^{N} \left(\sum_{k|n} 1 \right).$$

Inverting the orders of summation, we have

(2)
$$\sum_{n=1}^{N} d(n) = \sum_{k=1}^{N} \left[\frac{N}{k} \right],$$

a deceptively simple relation.
Thus we have

(3)
$$\sum_{n=1}^{N} d(n) = N \sum_{k=1}^{N} \frac{1}{k} + 0(N).$$

Using the fundamental relation of Euler, we have

(4)
$$\sum_{n=1}^{N} d(n) = N \log N + 0(N).$$

Exercises

1. Find the order of magnitude of $\Sigma_{n=1}^{N} d(an)$.

2. Find the order of magnitude of $\Sigma_{n=1}^{N} d(an + b)$.

3. Find the order of magnitude of $\Sigma_{n=1}^{N} \sigma_s(n)$.

4. Show that k and n/k are simultaneously divisors of n. These are distinct if n is not a perfect square.

5. Using this fact, obtain the order of magnitude of $\Sigma_{n=1}^{N} \sigma_s(n)$ with an error term $0(\sqrt{n})$.

4. A Simple Geometric Approach

We can do much better.
Consider the following diagram

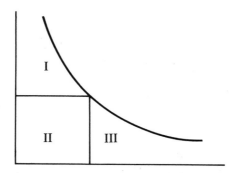

We know that the number of the lattice points determined by $1 \leqslant xy \leqslant n$ is given by the expression $\Sigma_{k=1}^{n} [n/k]$. From considerations of symmetry the number of lattice points in the region I is the same as in the region III. Hence, we have the relation

(1)
$$\sum_{k=1}^{n} \left[\frac{n}{k}\right] = 2 \sum_{k=1}^{\sqrt{n}} \left[\frac{n}{k}\right] + n$$

Hence, we have

(2)
$$\sum_{n=1}^{N} d(n) = N \log N + (2c - 1)N + 0(\sqrt{N}).$$

Exercises

1. Find the order of magnitude of $\Sigma_{n=1}^{N} (x/n - [x/n])$.

2. Use the same method to find an error term for $\Sigma_{n=1}^{N} r(n)$.

5. Analytic Equivalent

An analytic equivalent of the simpler geometric method presented above is the observation that k and n/k are simultaneously divisors of n. Hence, we have the relation, which we will use below,

(1)
$$d(n) = 2 \sum_{\substack{k \mid n \\ k \leqslant \sqrt{n}}} 1 - e(n),$$

where

(2)
$$e(n) = 1, \quad \text{if } n \text{ is a perfect square,}$$
$$= 0, \quad \text{otherwise.}$$

6. The Dirichlet Divisor Problem

Let us write

(1)
$$\Delta(N) = \sum_{n=1}^{N} d(n) - N \log N - (2c - 1)N,$$

Here C is the Euler Constant.

The Dirichlet divisor problem is the estimation of $\Delta(n)$. This is very difficult. Two methods are used at present. The first is due to Voronoi (Woronow). The second method uses trigonometric sums. Both involve complicated analysis.

In Chap. 12, we show that Tauberian theorems can be used to yield order of magnitude for the mean value of $\Delta(n)^2$.

The same problem for the sum $\Sigma_{n=1}^N d_k(n)$ is called the Piltz divisor problem. Here $d_k(n)$ is a higher divisor function.

7. The Mean Value of $d(n)^2$

We would expect from the foregoing that $d(n)$ behaves like $\log n$. To our surprise, this is not so. This means that occasionally, as we have pointed out $d(n)$ has very large values. To show this, let us find the mean value of $d(n)^2$.

We have

$$(1) \qquad \sum_{n=1}^N d(n)^2 = \sum_{n=1}^N d(n)\left(\sum_{k|n} 1\right).$$

Inverting the order of summation, and evaluating the sums that arise, we find that

$$(2) \qquad \sum_{n=1}^N d(n)^2 = C_1 N(\log N)^3 + 0(N(\log N)^2).$$

We leave the details to the reader.

Exercises

1. Use the generating function to obtain the foregoing results.

2. Evalute $\Sigma_{n=1}^N d(n)^3$.

3. Generally, evaluate $\Sigma_{n=1}^N d(n)^a$, where a is a positive integer. What happens if a is not?

4. Find the order of magnitude of $\Sigma_{n=1}^N \sigma_s(n)\sigma_s(n+a)$.

5. Evaluate the constant C_1 by two methods and equate.

8. Connection with Diophantine Equations

We observe that $d(n)$ is the number of solutions of $x_1 x_2 = n$. Consequently, the number of solutions of

(1) $$x_1 x_2 + x_3 x_4 = n$$

may be written as

(2) $$\sum_{k=1}^{N-1} d(k)d(n-k).$$

Similarly, the number of solutions of $x_1 x_2 - x_3 x_4 = n$, for $x_1 x_2 \leqslant N$ is equal to;

(3) $$\sum_{k=1}^{N-1} d(k)d(n+k).$$

We call sums of this type Ingham sums.

Exercises

1. Find the order of magnitude of the Ingham sum $\Sigma_{n=1}^{N} d(n)d(n+a)$.

2. Find the order of magnitude of $\Sigma_{n=1}^{N} d(n)d(N-n)$.

3. Find the order of magnitude of $\Sigma_{n=1}^{N} r(n)r(N-n)$.

4. Find the order of magnitude of $\Sigma_{n=1}^{N} r(n)r(n+a)$.

5. Consider the number of solutions of $x_1^2 + x_2^2 + x_3^2 + x_4^2 = n$. Show that this number may be written as $\Sigma_{n=0}^{N} r(n)r(N-n)$.

6. Evaluate the order of this sum, and hence prove Lagrange's theorem that every integer is the sum of four squares.

9. The Perron Sum Formula

Let us derive an expression for the partial sum of the coefficients of the Dirichlet series. This expression will be very useful, as we shall see in deriving various mean values. We start with the integral

(1) $$I(x) = \frac{1}{2\pi i} \int_C \frac{x^s ds}{s}.$$

Here C is a line parallel to the imaginary axis and lying to the right of it. We have

$$I(x) = 0, \quad 0 < x < 1,$$
$$= 1, \quad x > 1.$$

(2)

The first case is obtained by shifting C to the right; the second case is obtained by shifting C to the left and taking account of the residue at $s = 0$. Notice that we say nothing about the value for $x = 1$.

Using this integral, we have formally the result, where we assume x is not an integer,

(3)
$$\sum_{1 \leqslant n < x} a_n = \frac{1}{2\pi i} \int_C \frac{x^s}{s} f(s)\, ds,$$

Here C is a straight line to the right of the singularities of $f(s)$. Where

(4)
$$f(s) = \sum_{n=1}^{\infty} \frac{a_n}{n^s}.$$

It remains to prove this formula. We write

(5)
$$f(s) = \sum_{n=1}^{N} \frac{a_n}{n^s} + r_N(s)$$

where N is larger than x. Thus, we have

(6)
$$\frac{1}{2\pi i} \int_C \frac{f(s)x^s}{s}\, ds = \frac{1}{2\pi i} \int_C \frac{x^s}{s} \sum_{n=1}^{N} \frac{a_n}{n^s} + \frac{1}{2\pi i} \int_C \frac{x^s r_N(s)\, ds}{s}$$

$$= \sum_{1 \leqslant n < x} a_n + \frac{1}{2\pi i} \int_C \frac{x^s r_N(s)\, ds}{s}.$$

It remains to estimate the second term. To do this, we shall integrate by parts. We have

(7)
$$\int x^s r_N(s)\, ds = \sum_{n=N+1}^{\infty} \frac{a_n n^s}{\log(n/x)}.$$

We may integrate term-by-term if C is far enough to the right. If N is large enough, the term $\log(n/x)$ causes no difficulty. Thus, we see upon carrying out the integration by parts, that we can make the integral as small as desired by taking N sufficiently large. Thus, we obtain (1), the Perron sum formula.

Exercises

1. From the representation

$$\sum_{x < n < x+y} d(n) = \frac{1}{2\pi i} \int_C \left(\frac{(x+y)^s - x^s}{s} \right) \zeta^2(s)\, ds,$$

can we find a better estimate for the sum than the estimate given above?

2. Shift the contour, use the functional equation for $\zeta(s)$, and thus derive the representation of Voronoi for the error term.

3. Do the same for $r(n)$.

4. Do the same for $r_k(n)$.

5. Use the Perron sum formula for the interval $(x + y, x)$. By choosing y as a function of x, we can obtain a better result than by using the standard error term.

6. Show that the Perron sum formula can be obtained by taking the Mellin transform of the function $\sum_{n < x} \alpha_n$.

7. What do we get if we apply the Laplace transform?

8. Obtain a similar result for $\sum a_{mn}$, $m < x$, $n < y$.

10. Logarithmic Summability

It is easy to derive a generalization of this formula.
We begin with the representation

(1)
$$\frac{(\log x)^{k-1}}{(k-1)!} = \frac{1}{2\pi i} \int_C \frac{x^s\, ds}{s^k}$$

for any positive integer k. This can be proved in exactly the same way as the formula above. Using this formula, we have

$$\sum_{n \leqslant x} (\log x/n)^k a_n = \frac{k!}{2\pi i} \int_C \frac{x^s f(s)}{s^{k+1}} \, ds.$$

Exercises

1. Use this result to obtain an error term for the divisor function.

2. Use this result to obtain an error term for the circle function.

3. Use this result to obtain an error term for the higher divisor function.

11. Asymptotic Behavior

The Perron sum formula is used in the following way. We shift the contour to the left and determine the residue of the integrand. If we know the analytic behavior of $f(s)$, this is simple. This residue gives the order of magnitude of the partial sum. It remains to estimate the integral term. As we shall see, the mean value of the function $f(s)$ can be used for this. In general, we do not obtain a very sharp estimate in this way. However, we get the order of magnitude very easily and in Chap. 12; we shall see that this gives us an estimate for the mean value of the square of the error term.

Let us now give an example of how the sum formula can be used. Consider the expression.

(1)
$$\sum_{1 \leqslant n < x} d(n) = \frac{1}{2\pi i} \int_C \frac{x^s \zeta^2(s) \, ds}{s}.$$

Here C is a line lying to the right of the line $\sigma = 1$. Shifting the contour to the left, we pass a single pole at $s = 1$.

Thus we have the expression

(2)
$$\Delta(x) = \frac{1}{2\pi i} \int_{C'} \frac{x^s \zeta^2(s) \, ds}{s}.$$

Here C' is a line to the left of $\sigma = 1$. In order to make this integral as small as possible, we would like to shift as far as possible to the left. To estimate this remainder term, it is convenient to use a contour of the following form

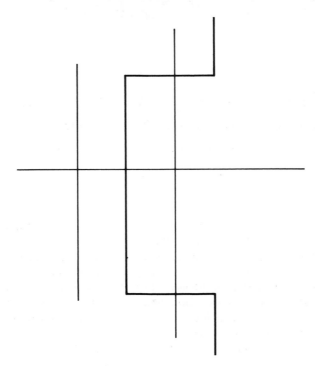

It remains to estimate the integral along the various parts. The integral over the lines parallel to the x-axis are easily estimated. The integral along the line to the left of the residue at $s = 1$ is estimated by using a mean value result for $\zeta(s)$.
We have

(3)
$$\left| \int \frac{x^s \zeta(s)\, ds}{s} \right| \leqslant x^\sigma \int \left| \frac{\zeta^2(s)}{s} \right| ds.$$

The integral here is taken over this vertical line. Using the Cauchy-Schwarz inequality we have that this term is bounded by the expression

(4)
$$\left(\int \frac{|\zeta(s)|^4 ds}{|s|} \right)^{\frac{1}{2}} \left(\int \frac{ds}{|s|} \right)^{\frac{1}{2}}.$$

Thus, we see that our ability to estimate the error term depends upon our ability to estimate

(5) $$\int |\zeta(s)|^4 ds.$$

Thus we see how the estimate of the error term can be made to depend on the estimation for various mean values. In Chap. 12, we will pursue this subject further.

12. The Mean Value of $d(an^2 + bn + c)$

Using the representation of $d(n)$ given above, in (1) of Sec. 5, we have

(1) $$\sum_{n=1}^{N} d(an^2 + bn + c) = 2 \sum_{n=1}^{N} \left(\sum_{\substack{k/(an^2+bn+c) \\ k \leqslant \sqrt{|an^2+bn+c|}}} 1 \right)$$

Inverting the order of summation and using some algebraic number theory, we have the estimate,

(2) $$\sum_{n=1}^{N} d(an^2 + bn + c) = f(a,b,c)\, N (\log N) + 0(N)$$

Exercises

1. Find the order of magnitude of $\Sigma_{n=1}^{N} d(n^2)$.

2. Find the order of magnitude of $\Sigma_{n=1}^{N} \sigma_s(an^2 + bn + c)$.

13. The Mean Value of $d(p(n))$

We assume, as usual, that $p(n)$ is irreducible.

If we try the same method for the mean value for $d(p(n))$, we find that the error estimate overwhelms what is supposed to be the principle term. We conjecture that

(1) $$\sum_{n=1}^{N} d(p(n)) = c_2 N \log N + 0(N),$$

where the constant c_2 depends upon the polynomial $p(n)$.

14. Ramanujan Expansions

If we wish to obtain more precise results, it appears that we must use a different method. Fortunately, we have expansions which permit us to obtain more refined mean value estimates. Such expansions were given by Ramanujan. Thus, we have,

(1)
$$\sigma_{-s}(n) = \zeta(1+s) \sum_{q=1}^{\infty} \frac{c_q(n)}{q^s},$$

$\text{Re}(s) > 0$, where, as usual $\sigma_{-s}(n) = \sum_{k \mid n} 1/k^s$.

Using some estimates for trigonometric sums various mean values can be obtained from this expression. Reference is given at the end of the chapter.

Exercises

1. Obtain the foregoing expansion by decomposing the function

$$\sum_{n=1}^{\infty} \frac{x^n}{n^s(1-x^n)}$$

into partial fractions.

2. Obtain an expansion for $\Pi_{n=1}^{\infty} (1 + x^n)$.

Miscellaneous Exercises

1. Show that every number is uniquely representable by powers of two.

2. Show that an analytic equivalent of this statement is the identity

$$\frac{1}{1-x} = (1+x)(1+x^2)(1+x^4)\cdots.$$

3. Establish this identity by means of the identity

$$\frac{1-x^{2^N}}{1-x} = (1+x)(1+x^2)\cdots(1+x^{2^{N-1}}).$$

4. Let $\alpha(n)$ be the number of ones in the representation of n. Find the order of magnitude of $\Sigma_{n=1}^{N}\,\alpha(n)$.

5. Find the order of magnitude of $\Sigma_{n=1}^{N}\,\alpha(\alpha(n))$. (See

Bellman, R., and H. N. Shapiro, *On a Problem in Additive Number Theory,* Ann. Math. **49** (1948), 333–340.

Grosswald, E., *Properties of Some Arithmetical Function,* J. Math. Anal. Appl. **28** (1969), 405–430.

Stewart, B. M., *Sums of Functions of Digits,* Canad. J. Math. **12** (1960), 374–389.

Mahler, K., *On the Generating Functions of Integers with a Missing Digit, K'o Hsüeh* (Science) **29** (1947), 265–267.

Alexander, R., *Remarks About the Digits of Integers,* J. Austral. Math. Soc. **12** (1971), 239–241.

Thorp, E., and R. Whitley, *Poincaré's Conjecture and the Distribution of Digits in Tables,* Compositio Math. **23** (1971), 233–250.

6. Can one find the order of magnitude of $\Sigma_{n=1}^{N}\,d(d(n))$?

7. Show that

$$\sum_{n=1}^{\infty}\frac{1}{d(n)n^s} = \zeta(s)^{1/2}g(s),$$

where $g(s)$ has an absolutely convergent Dirichlet series for $\mathrm{RE}(s) > \frac{1}{2}$.

Bibliography and Comments

Section 1

The problem of the number of solutions of $d(n) = k$ is very interesting. See

Erdös, P., and L. Mirsky, *The Distribution of Values of the Divisor Function* d(n), Proc. London Math. Soc. 2 (3) (1952), 257–271.

Rankin, R. A., *The Divisibility of Divisor Functions,* Proc. Glasgow Math. Assoc. 5 (1961), 35–40.

Erdös, P., *Some Remarks on the Functions φ and σ,* Bull. Acad. Polon. Sci. Ser. Sci. Math. Astronom. Phys. 10 (1962), 617–619.

Spira, R., *The Complex Sum of Divisors,* Amer. Math. Monthly 68 (1961), 120–124.

Cohen, E., *Arithmetical Notes.* IV, *A Set of Integers Related to the Divisor Function,* J. Tennessee Acad. Sci. 37 (1962), 119–120.

——, *Arithmetical Notes.* XI, *Some Divisor Identities,* Enseignement Math. 10 (2) (1964), 248–254.

Titchmarsh, E. C., *On Divisor Problems,* Quart. J. Math. Oxford Ser. 9 (1938), 216.

Section 2

See

Harris, V. C., and J. J. Warren, *A Generating Function for $\sigma_k(n)$,* Amer. Math. Monthly 66 (1959), 467–472.

Cohen, E., *Arithmetical Notes.* V, *A Divisibility Property of the Divisor Function,* Amer. J. Math. 83 (1961), 693–697.

Kesava, M. P., *Some Generalizations of the Divisor Function,* J. Indian Math. Soc. (N.S.), 9 (1945), 32–36.

Section 3

See

Chowla, S., *Note on a Certain Arithmetical Sum,* Proc. Nat. Ins. Sci. India 13 (5) (1947), 1.

Titchmarsh, E. C., *On Series Involving Divisors,* J. London Math. Soc. 22 (1947), 179–184.

Halberstam, H., *An Asymptotic Formula in the Theory of Numbers,* Trans. Amer. Math. Soc. 84 (1957), 338–351.

Atkinson, F. V., *A Divisor Problem,* Quart. J. Math. Oxford Ser. 12 (1941), 193–200.

Titchmarsh, E. C., *Some Problems in the Analytic Theory of Numbers,* Quart. J. Math. Oxford Ser. 13 (1942), 129–152.

Bush, L. E., *An Asymptotic Formula for the Average Sum of the Digits of Integers,* Amer. Math. Monthly **47** (1940), 154-156.

Dressler, R. E., *An Elementary Proof of a Theorem of Erdos on the Sum of Divisors Function,* J. Number Theory **4** (1972), 532-536.

Wall, C. R., P. L. Johnson, and B. Donald, *Density Bounds for the Sum of Divisors Function,* Math. Comp. **26** (1972), 773-777.

Wells, C. R., *Density Bounds for the Sum of Divisors Functions, The Theory of Arithmetic Functions* (Proc. Conf., Western Michigan University, Kalamazoo, Michigan, 1971), pp. 283-287. Lecture Notes in Math. Vol. 251, Springer, Berlin, 1972.

Redish, K. A., *Tables of Sequences of Divisor-Sums,* Proc. of the 25th Summer Meeting of the Canadian Mathematical Congress (Lakehead University, Thunder Bay, Ontario, 1971), pp. 527-537, Lakehead University, Thunder Bay, Ontario, 1971.

We have not discussed the difficult problem of the mean value over small intervals. For some results in that direction see

Bellman, R., and H. N. Shapiro, *On the Normal Order of Arithmetic Functions,* Proc. Nat. Acad. Sci. **38** (10) (1952), 884-886.

See also

Chowla, S. and H. Walum, *On the Divisor Problem,* Proc. Sympos. Pure Math. Vol. VIII, pp. 138-143, Amer. Math. Soc., Providence, Rhode Island, 1965.

Dixon, R. D., *On a Generalized Divisor Problem,* J. Indian Math. Soc. (N.S.) **28** (1964), 187-196.

Rodriquez, G., *Sul Problema dei Divisori di Titchmarsh,* Boll. Un. Mat. Ital. **20** (3) (1965), 358-366.

Scholomiti, N. C., *A Property of the τ-Function,* Amer. Math. Monthly **72** (1965), 745-47.

Duncan, R. L., *Some Estimates for* $\sigma(n)$, Amer. Math. Monthly **74** (1967), 713-715.

Horadam, E. M., *A Sum of a Certain Divisor Function for Arithmetical Semi-Groups,* Pacific J. Math. **22** (1967), 407-412.

Erdös, P. and I. Kátai, *On the Sum* $\Sigma d_4(n)$, Acta Sci. Math. (Szeged) **30** (1969), 313-324.

Krätzel, E., *Teilerprobleme in Drei Dimensionen,* Math. Nachr. **42** (1969), 275-288.

Erdös, P., *On the Sum* $\Sigma_{n=1}^{x} d[d(n)]$, Math. Student **36** (1968), 227-229.

Mientka, W. E. and R. L. Vogt, *Computational Results Relating to Problems Concerning* $\sigma(n)$, Mat. Vesnik, **22** (7) (1970), 35-36.

Motohashi, Y., *On the Sum of the Number of Divisors in a Short-Segment,* Acta Arith. **17** (1970), 249-253.

Delmer, F., *Sur la Somme de Diviseurs* $\Sigma_k \leqslant x\{d\,[f(k)]\}$, C.R. Acad. Sci. Paris Sér. A-B **272** (1971), A849–A852.

Berndt, B. C., *On the Average Order of a Class of Arithmetical Functions.* I, J. Number Theory **3** (1971), 184–203.

Webb, W. A., *An Asymptotic Estimate for a Class of Divisors Sums,* Duke Math. J. **38** (1971), 575–582.

Porter, J. W., *The Generalized Titchmarsh-Linnik Divisor Problem,* Proc. London Math. Soc. **24** (3) (1977), 15–26.

Section 5

For a discussion of the mean value of $d(n)$ over small intervals, see

Chih, Tsung-tao, *A Divisor Problem,* Acad. Sinica Sci. Record, **3** (1950), 177–182.

Section 6

See the paper by I. M. Vinogradov cited in Chap. 3.
See also

Davenport, H., *A Divisor Problem,* Quart. J. Math. Oxford Ser. **20** (1949), 37–44.

Beumer, M. G., *The Arithmetical Function* $\tau_k(N)$, Amer. Math. Monthly **69** (1962), 777–781.

Yüh, Ming I., and F. Wu, *On the Divisor Problem for* $d_3(n)$, Sci. Sinica **11** (1962), 1055–1060.

Chowla, S., and H. Walum, *On the Divisor Problem,* Norske Vid. Selsk. Forh. (Trondheim) **36** (1963), 127–135.

Cleaver, F. L., *On a Theorem of Voronoi,* Trans. Amer. Math. Soc. **120** (1965), 390–400.

Segal, S. L., *A Note on the Average Order of Number-Theoretic Error Terms,* Duke Math. J. **32** (1965), 279–284.

Katai, I., *On Oscillations of Number-Theoretic Functions,* Acta Arith. **13** (1967/68), 107–122.

Diamond, H. G., *Interpolation of the Dirichlet Divisor Problem,* Acta Arith. **13** (1967/68), 151–168.

For a detailed discussion of the Dirichlet divisor problem and the Piltz divisor problem, see the book

Chandrasekharan, K., *Arithmetical Functions,* Springer-Verlag, New York, 1970.

Section 7

See

Smith, R. A., *On* Σr(n)r(n + a), Proc. Nat. Inst. Sci. India Part A, **34** (1968), 132–137.

Section 8

See

Johnson, S. M., *On the Representations of an Integer as the Sum of Products of Integers,* Trans. Amer. Math. Soc. **76** (1954), 177–189.
Bellman, R., *On Some Divisor Sums Associated with Diophantine Equations,* Quart. J. Math. Oxford Ser. **1** (1950), 136–146.
Hooley, C., *An Asymptotic Formula in the Theory of Numbers,* Proc. London Math. Soc. **7** (3) (1957), 396–413.

Section 13

See

Erdös, P., *On Some Problems of Bellman and a Theorem of Romanoff,* J. Chinese Math. Soc. (N.S.) **1** (1951), 409–421 (Chinese).

Section 14

The use of Ramanujan sums involves trigonometric sums. For further information, see the papers

Bellman, Richard, *On the Average Value of Arithmetic Functions,* Proc. Nat. Acad. Sci. U.S.A. **34** (1948), 149–152.
Bellman, Richard and Harold N. Shapiro, *On the Sum* Σd(an² + bn + c) *and related sums* (unpublished).
Bellman, Richard, *Ramanujan Sums and the Average Value of Arithmetic Functions,* Duke Math. J. **17** (1950), 159–168.
Carlitz, L., *On a Problem in Additive Arithmetic.* II, Quart. J. Math. Oxford Ser. **3** (1932), 273–290.
Carmichael, R. D., *Expansions of Arithmetical Functions in Infinite Series,* Proc. London Math. Soc. **34** (2) (1932), 1–26.
Cohen, Eckford, *An Extension of Ramanujan's Sum,* Duke Math. J. **16** (1949), 85–90.
Gordon, B. and K. Rogers, *Sums of the Divisor Function,* Canad. J. Math. **16** (1964), 151–158.
Hardy, G. H., *Note on Ramanujan's Trigonometrical Function* $c_q(n)$, *and Certain Series of Arithmetical Functions,* Proc. Cambridge Philos. Soc. **20** (1921), 263–271.
Holder, O., *Zur Theorie der Kreisteilungsgleichung* $K_m(x) = 0$, Prace Mat. Fiz. **43** (1936), 13–23.

Hua, L. K., *On an Exponential Sum,* J. Chinese Math. Soc. **2** (1940), 301–312.

Kac, M., E. R. Van Kampen, and Aurel Wintner, *Ramanujan Sums and Almost Periodic Functions,* Amer. J. Math. **62** (1940), 107–114.

Ramanujan, S., *On Certain Trigonometrical Sums and Their Applications in the Theory of Numbers,* Trans. Cambridge Philos. Soc. **22** (1918), 259–276.

Suryanarayana, D., and R. Sita Rama Chandra Rao, *On an Asymptotic Formula of Ramanujan,* Math. Scand. **32** (1973), 258–264.

Titchmarsh, E. C., *A Divisor Problem,* Rend. Circ. Mat. Palermo **54** (1930), 414–429.

van der Corput, J. G., *Une unégalité rélative au nombre des diviseurs,* Nederl. Akad. Wetensch. Proc. Ser. **42** (1939), 547–553.

Walfisz, Arnold, *Additive Zahlentheorie.* II, Math. Z. **40** (1936), 592–607.

10.   The Squarefree Problem

1. Introduction

In this chapter, we want to study some aspects of squarefree numbers.

Squarefree numbers are numbers which possess no repeated prime factors. We would like to solve the corresponding problems for primes. What we hope is that consideration of squarefree numbers will give us some ideas of the order of difficulties of the corresponding problems for primes. What we find is that even the problems for squarefree numbers are quite difficult.

In the second section, we find a representation which makes the mean value of squarefree numbers quite easy to obtain. When we try to extend this method to numbers of the form $n^2 + 1$, we find, as usual, that the estimate of the error term causes difficulty. We can circumvent this difficulty by considering the Pell equation, an interesting equation in its own right. Then we turn to polynomials of higher degree. In the general case, all we can do is make a conjecture. The method used here is too weak. What we would like to prove is that there are infinitely many squarefree numbers of the form $n^2 + 1$ without finding a mean value.

We would like to consider the general quadratic polynomial $an^2 + bn + c$. We consider the polynomial $n^2 + 1$ in order that we can use the known properties of the Pell equation. This simplifies the proof.

2. The Generating Function

We have

$$\sum_{n=1}^{\infty} \frac{\mu^2(n)}{n^s} = \prod_{p}(1 + 1/p^s)$$

Richard Bellman, Analytic Number Theory: An Introduction   ISBN 0-8053-0452-5

(1)
$$= \prod_{p} \left(\frac{1 - 1/p^{2s}}{1 - 1/p^s} \right)$$

$$= \zeta(s)/\zeta(2s).$$

From this, we obtain the fundamental representation

(2)
$$\mu^2(n) = \sum_{k^2 \mid n} \mu(k).$$

Exercises

1. Obtain this representation by argument without using a generating function.

2. Show that the generating function of the rth power free integers is given by $\zeta(s)/\zeta(rs)$.

3. The Mean Value of $\mu^2(n)$

We have

$$\sum_{n=1}^{N} \mu^2(n) = \sum_{n=1}^{N} \left(\sum_{k^2 \mid n} \mu(k) \right)$$

$$= \sum_{k=1}^{\sqrt{N}} \mu(k) \left(\sum_{\substack{n \equiv 0(k^2) \\ 1 \leqslant n < N}} 1 \right)$$

(1)
$$= \sum_{k=1}^{\sqrt{N}} \mu(k) \left[\frac{N}{k^2} \right]$$

$$= N \sum_{k=1}^{\infty} \frac{\mu(k)}{k^2} + 0(\sqrt{N})$$

$$= \frac{6N}{\Pi^2} + 0(\sqrt{N})$$

Exercises

1. Obtain a corresponding result for rth power free integers.

2. Improve the error estimate by assumption of the Riemann hypothesis.

3. From the representation

$$\sum_{x<n<x+y} \mu^2(n) = \frac{1}{2\pi i} \int_C \left(\frac{(x+y)^s - x^s}{s} \right) \frac{\zeta(s)\,ds}{\zeta(2s)} ,$$

can we find a better estimate for the sum than using the estimate of the error term above?

4. The Mean Value of $\mu^2(n^2 + 1)$

If we use the same technique as above, we find we have difficulty, as usual, with the error estimate. Consequently, we must employ a new technique.

We shall study the Pell equation to obtain an estimate for the number of solutions of

(1) $$n^2 + 1 \equiv 0(y^2)$$

for $n \leqslant x$ and y sufficiently large.

5. The Pell Equation

The congruence above is equivalent to the equation

(1) $$n^2 + 1 = ky^2 .$$

The equation

(2) $$n^2 - km^2 = -1$$

is known as the Pell equation, a very interesting equation in its own right. Let us study some of the properties of the solutions of this equation.

We associate with every solution the number $n + m\sqrt{k}$. We have the basic factorization

(3) $$n^2 - km^2 = (n - m\sqrt{k})(n + m\sqrt{k}),$$

which shows the genesis of the association above.

It follows from this factorization that the product of two solution numbers yields another solution number. Even further, the quotient of two solution numbers yields another solution number. What we want to show is that all solutions are generated by one solution.

To do this, we proceed as follows. Associate, as we have said before, with the solution the number $n + m \sqrt{k}$. If n and m are positive this quantity is greater than one. Let $n_1 + m_1 \sqrt{k}$ be the smallest positive solution greater than one. Suppose that there is another solution $n_2 + m_2 \sqrt{k}$ which is not a power of the first. Then it must be between two powers of the first.

By this we mean

(4) $$(n_1 + m_1 \sqrt{k})^a < n_2 + m_2 \sqrt{k} < (n_1 + m_1 \sqrt{k})^{a+1}.$$

Consequently, we have

(5) $$1 < \frac{n_2 + m_2 \sqrt{k}}{(n_1 + m_1 \sqrt{k})^a} < n_1 + m_1 \sqrt{k}.$$

From what we have said before, the quantity $(n_2 + m_2\sqrt{k})/(n_1 + m_1 \sqrt{k})^a$ represents a solution of the Pell equation. On the other hand we see from Eq. (5) that it lies between one and $n_1 + m_1 \sqrt{k}$, a contradiction. Consequently, all solutions of the Pell equation are powers of the solution equivalent to $n_1 + m_1 \sqrt{k}$.

Exercises

1. Show that all the numbers equivalent to the solution of the Pell equation form a group with a single generator.

2. Associate with the solutions of $x^3 + ky^3 + k^2z^3 - 3kxyz = 1$, the number $x + \sqrt[3]{k}\, y + \sqrt[3]{k^2}z$. Show that these numbers form a group.

3. Does this group have a single generator?

6. The Number of Squarefree Numbers of the Form $n^2 + 1$

We are now ready to discuss the number of squarefree numbers of the form $n^2 + 1$. The congruence we obtain leads to the equation $n^2 + 1 = ky^2$. We could, if we wanted to, use the foregoing methods to discuss this equation. Instead, let us use the fact that the quotient of two solutions of this equation is a solution of the Pell equation. If we consider the interval $1 \leqslant n \leqslant x,\ 1 \leqslant y \leqslant x$, we see that for fixed k the number of solutions in this interval is $O(\log x)$. Consequently, to estimate the sum $\Sigma_{n=1}^{N} \mu^2(n^2 + 1)$, we proceed as follows. We write

(1) $$\sum_{n=1}^{N} \mu^2(n^2 + 1) = \sum_{n=1}^{N} \left(\sum_{k^2|(n^2+1)} \mu(k) \right).$$

Inverting the order of summation, we break up the resulting sum into two parts. Thus, we have

(2)
$$\sum_{n=1}^{N^2} = \sum_{k=1}^{N} + \sum_{k=N^{2/3}+1}^{N^2}.$$

In the first sum we proceed as before. To estimate the second sum we use the fact that the number of solutions of $n^2 + 1 = ky^2$ in the interval $1 \leqslant n \leqslant N$, $1 \leqslant k \leqslant N$ is $0 (\log N)$ for each fixed y. Consequently, if y is in the interval $k \geqslant N^{2/3} + 1$, we see that k must be in the interval $1 \leqslant y \leqslant N^{2/3}$. We see that the sum is at most $0(N^{2/3} \log N)$. Consequently, we have the result

(3)
$$\sum_{n=1}^{N} \mu^2 (n^2 + 1) = c_1 N + 0(N^{2/3} \log N).$$

The equation need not have a solution for every k. Obviously, it is not necessary to consider these equations.

A very simple sieving argument due to H. N. Shapiro works for the case of $n^2 + 1$. Namely: let x be an integer $\geqslant 4$, $Q(x)$ = number of $n^2 + 1$, $n = 1, 2, \ldots,$ x which are squarefree, $B(p^2)$ = the number of solutions mod p^2 to $n^2 + 1 \equiv 0 \pmod{p^2}$, p a prime. Then we know that $B(2^2) = 0$, and for p an odd prime, $B(p^2) = 0$ for $p \equiv -1 \pmod 4$ and $B(p^2) = 2$ for $p \equiv 1 \pmod 4$. Then

(4)
$$Q(x) \geqslant x - \sum_{p \leqslant x} (\text{no. of } n < x \text{ s.t. } n^2 + 1 \equiv 0 \pmod{p^2}))$$

so that

(5)
$$Q(x) \geqslant x - \sum_{p \leqslant \sqrt{x}} \frac{x}{p^2} B(p^2) - 2 \sum_{\sqrt{x} < p \leqslant x}.$$

But using the information on $B(p^2)$ we have

(6)
$$\sum_{p \leqslant \sqrt{x}} \frac{B(p^2)}{p^2} < \sum_{n \geqslant 5} \frac{2}{r^2} < \frac{1}{2},$$

and from elementary prime number theory $2 \sum_{p \leqslant x} 1 < c(x/\log x)$, so that (2) yields

(7) $$Q(x) > x\left(\frac{1}{2} - \frac{c}{\log x}\right) > \frac{x}{3} \quad \text{(for } x \text{ large)},$$

and this certainly implies infinitely many squarefrees of the form $n^2 + 1$.

This simple argument can be elaborated upon so as to give an asymptotic result. Namely, let $L = L(x)$ to be chosen later, and let $\Delta = \Pi_{p \leqslant L} p$. Then

(8) $Q(x)$ = (number of $n \leqslant x$ such that $n^2 + 1$ is not divisible by any p^2, p a prime, $p \leqslant L$) = $\hat{Q}(x)$

$$+ 0\left(x \sum_{p > L} \frac{1}{p^2}\right) + 0\left(\frac{x}{\log x}\right).$$

But

(9) $$\hat{Q}(x) = \sum_{\substack{n \leqslant x \\ k | \Delta \\ k^2 / n^2 + 1}} \sum \mu(k) = \sum_{k | \Delta} \mu(k) \left\{\frac{x}{k^2} B(k^2) + 0(B(k^2))\right\}$$

$$= x \prod_{p | \Delta} \left(1 - \frac{B(p^2)}{p^2}\right) + 0\left(\sum_{k | \Delta} 2^{2(k)}\right)$$

$$= x \prod_{p | \Delta} \left(1 - \frac{B(p^2)}{p^2}\right) + 0\left(3^{\pi(\Delta)}\right).$$

Then it is easy to see that

$$\prod_{p | \Delta} \left(1 - \frac{B(p^2)}{p^2}\right) = \prod_p \left(1 - \frac{B(p^2)}{p^2}\right) + 0\left(\frac{\log L}{L}\right)$$

so that putting everything together

(10) $$Q(x) = \prod_p \left(1 - \frac{B(p^2)}{p^2}\right) + 0\left(\frac{\log L}{L}\right) + 0\left(\frac{c^{L/\log L}}{x}\right).$$

Thus taking $L = c_1 (\log x)(\log \log x)$, (for small $c_1 > 0$), we get

(11) $$\frac{Q(x)}{x} = \prod_p \left(1 - \frac{B(p^2)}{p^2}\right) + 0 \frac{1}{\log x}.$$

Exercise

1. Find the order of magnitude of $\Sigma_{n=1}^{N} \mu^2(an^2 + bn + c)$.

7. The Number of Squarefree Numbers in a General Polynomial Sequence

If we try to employ the foregoing method for the general case, we encounter difficulties. Consider, for example, the case where we have $n^3 + 2$. We know, from Thue-Siegel-Roth theorem, that the equation $n^3 + 2 = ky^2$ has only a finite number of solutions. However, at present, we have little information about what values of k have a solution at all, and how to estimate the number of solutions in this case. The method of Roth can be used to estimate the number of solutions but this estimate is too weak to use.

Using difficult techniques, particular polynomials can be treated.

It is possible that a method based on trigonometric sums will provide the information we require. In any case, we conjecture that if $p(n)$ is irreducible, we have

$$(1) \qquad \sum_{n=1}^{N} \mu^2(p(n)) = c_2 N + 0(N),$$

where c_2 is a constant which depends upon $p(n)$.

The correct condition is not that $p(n)$ be irreducible but that there does not exist a prime q such that q^2 divides $p(n)$ for all n. Then one conjectures (1) with a positive constant c_2. (Note that an irreducible polynomial can be divisible by a square factor for all integer values of argument; e.g., $x^4 - x^2 + 4$ is always divisible by 4.)

If $f(x) = \Pi_{i=1}^{r} g_i(x)$ where each $g_i(x)$ is either a linear or quadratic polynomial, then for $x = 1, \cdots, N$ the number of squarefree $f(x)$ is $cx + 0(x)$ for some $c > 0$.

Miscellaneous Exercises

1. Find the density of integers n where n and $n + 2$ are both squarefree.

2. Find the density of squarefree numbers in an arithmetic progression. (Landau)

Bibliography and Comments

Section 1

The study of squarefree integers naturally leads to the study of kth power free integers. See

Atkinson, F. V., and Cherwell, *The Mean-Value of Arithmetical Functions,* Quart. J. Math. Oxford Ser. **20** (1949), 65-79.

Cohen, E., *The Average Order of Certain Types of Arithmetical Functions: Generalized* k-*Free Numbers and Totient Points,* Monatsh. Math. **64** (1960), 251-262.

——, *The Number of Representations of an Integer as a Sum of Two Squarefree Numbers,* Duke Math. J. **32** (1965), 181-185.

——, *Some Sets of Integers Related to the* k-*Free Integers,* Acta Sci. Math. (Szeged) **22** (1961), 223-233.

——, *On the Distribution of Certain Sequences of Integers,* Amer. Math. Monthly **70** (1963), 516-521.

——, *An Elementary Estimate for the* k-*Free Integers,* Bull. Amer. Math. Soc. **69** (1963), 762-765.

——, *Arithmetical Notes.* X, *A Class of Totients,* Proc. Amer. Math. Soc. **15** (1964), 534-539.

——, *Some Asymptotic Formulas in the Theory of Numbers,* Trans. Amer. Math. Soc. **112** (1964), 214-227.

Evelyn, C. J. A., *Relations Between Arithmetical Functions,* Proc. Cambridge Philos. Soc. **63** (1967), 1027-1029.

Gioia, A. A., and A. M. Vaidya, *The Number of Squarefree Divisors of an Integer,* Duke Math. J. **33** (1966), 797-799.

Harris, V. C., and M. V. Subbarao, *An Arithmetic Sum with an Application to Quasi* k-*Free Integers,* J. Austral. Math. Soc. **15** (1973), 272-278.

Joshi, V. S., *On the Order of Some Error Functions Related to* K-*Free Integers,* Proc. Amer. Math. Soc. **35** (1972), 325-332.

McCarthy, P. J., *On a Certain Family of Arithmetic Functions,* Amer. Math. Monthly **65** (1958), 586-590.

Mirsky, L., *A Property of Squarefree Integers,* J. Indian Math. Soc. (N.S.) **13** (1949), 1-3.

Nymann, J. E., *A Note Concerning the Square-Free Integers,* Amer. Math. Monthly **79** (1972), 63-65.

Pakshirajan, R. P., *Some Properties of the Class of Arithmetic Functions* $T_r(N)$, Ann. Polon. Math. **13** (1963), 113-114.

Rogers, K., *The Schnirelmann Density of the Squarefree Integers,* Proc. Amer. Math. Soc. **15** (1964), 515-516.

Roth, K. F., *A Theorem Involving Squarefree Numbers,* J. London Math. Soc. **22** (1947), 231-237.

Sklar, A., *On the Factorization of Squarefree Integers,* Proc. Amer. Math. Soc. **3** (1952), 701–705.

Subbarao, M. V., and Y. K. Feng, *On the Distribution of Generalized K-Free Integers in Residue Classes,* Duke Math. J. **38** (1971), 741–748.

Subbarao, M. V., and D. Suryanarayana, *On the Order of the Error Functions of the* (k, r)-*Integers,* J. Number Theory **6** (1974), 112–123.

Suryanarayana, D., and R. Sitaramachandrarao, *The Number of Square-Full Divisors of an Integer,* Proc. Amer. Math. Soc. **34** (1972), 79–80.

Suryanarayana, D., *Semi-k-Free Integers,* Elem. Math. **26** (1971), 39–40.

——, *A Remark on 'Uniform 0-Estimates of Certain Error Functions Connected with k-Free Integers,'* J. Austral. Math. Soc. **15** (1973), 177–178.

Suryanarayana, D., and V. S. R. Prasad, *The Number of k-Free Divisors of an Integer,* Acta Arith. **17** (1970/71), 345–354.

Tull, J. P., *Average Order of Arithmetic Functions,* Illinois J. Math. **5** (1961), 175–181.

Wagstaff, S. S., Jr., *On k-Free Sequences of Integers,* Math. Comp. **26** (1972), 767–771.

The difference between primes leads to quite difficult analysis. Even the difference between consecutive squarefree integers requires difficult analysis.

Halberstein, H. and K. F. Roth, *On the Gaps Between Consecutive k-Free Integers,* J. London Math. Soc. **26** (1951), 268–273.

Richert, H-E., *On the Difference Between Consecutive Squarefree Numbers,* J. London Math. Soc. **29** (1954), 16–20.

Roth, K. F., *On the Gaps Between Squarefree Numbers,* J. London Math. Soc. **26** (1951), 263–268.

Yüh, Ming-I., *On the Difference Between Squarefree Numbers,* Sci. Record (N.S.) **1** (3) (1957), 13–16.

Again, we have avoided the question of the distribution over small intervals. See

Bellman, R., and H. N. Shapiro, *The Distribution of Squarefree Integers in Small Intervals,* Duke Math. J. **21** (1954), 629–638.

Kátai, I., *On the Values of Multiplicative Functions in Short Intervals,* Math. Ann. **183** (1969), 181–184.

Section 2

See

Halberstam, H., and H.-E. Richert, *Mean Value Theorems for a Class of Arithmetic Functions,* Acta Arith. **18** (1971), 243–256.

Hooley, C., *On the Distribution of Square-Free Numbers,* Canad. J. Math. **25** (1973), 1216–1223.

Orr, R. C., *Remainder Estimates for Squarefree Integers in Arithmetic Progression*, J. Number Theory **3** (1971), 474–497.

Section 3

See

McCarthy, P. J., *The Probability that* (n, f(n)) *is r-Free*, Amer. Math. Monthly **67** (1960), 368–369.

For *k*th power results, see

Fluch, W., *Bemerkung über Quadratfreie Zahlen in Arithmetischen Progressionen*, Montash. Math. **72** (1968), 472–430.

McCarthy, P. J., *Some Remarks on Arithmetical Identities*, Amer. Math. Monthly **67** (1960), 539–548.

Mirsky, L., *Arithmetical Pattern Problems Relating to Divisibility by* r-th *Powers*, Proc. London Math. Soc. **50** (2) (1949), 497–508.

Moser, L., and R. A. MacLeod, *The Error Term for the Squarefree Integers*, Canad. Math. Bull. **9** (1966), 303–306.

Onishi, H., *The Number of Positive Integers* $n \leqslant N$ *such that* n, n + a_2, N + a_3, $\cdots$, n + a_r *are all Square-Free*, J. London Math. Soc. **41** (1966), 138–140.

Stark, H. M., *On the Asymptotic Density of* k-*Free Integers*, Proc. Amer. Math. Soc. **17** (1966), 1211–1214.

Suryanarayana, D., and R. Sitaramachandra Rao, *On the Order of the Error Function of the* k-*Free Integers*, Proc. Amer. Math. Soc. **28** (1971), 53–58.

Suryanarayana, D., *Uniform 0-Estimates of Certain Error Functions Connected with* k-*Free Integers*, J. Austral. Math. Soc. **11** (1970), 242–250.

——, *The Number and Sum of* k-*Free Integers* $\leqslant$ x *which are Prime to* n, Indian J. Math. **11** (1969), 131–139.

Vaidya, A. M., *On the Changes of Sign of a Certain Error Function Connected with* k-*Free Integers*, J. Indian Math. Soc. (N.S.) **32** (1968), 105–111.

——, *On the Order of the Error Function of Square-Free Numbers*, Proc. Nat. Inst. Sci. India Sect. A **32** (1966), 196–201.

Warlimont, R., *On Squarefree Numbers in Arithmetic Progression*, Monatsh. Math. **73** (1969), 433–448.

——, *On Divisor Problems in Connection with Squarefree Numbers*, Montash. Math. **74** (1970), 154–165.

Section 5

See

Copley, G. N., *Recurrence Relations for the Solutions of Pell's Equation*, Amer. Math. Monthly **66** (1959), 288–290.

The Pell equation is one of the most interesting Diophantine equations. For an account of this equation, see the books

Dickson, L. E., *History of the Theory of Numbers,* Chelsea, New York, 1952.

Hardy, G. H. and E. M. Wright, *An Introduction to the Theory of Numbers,* Clarendon Press, Oxford, 1960.

Section 7

See

Grosswald, E., *On Some Conjectures of Hardy and Littlewood,* Publ. Ramanujan Inst. **1** (1968/69), 75–89.

Lehmer, D. H. and E. Lehmer, *Heuristics, Anyone? Studies in Mathematical Analysis and Related Topics,* pp. 202–210, Stanford Univ. Press, Stanford, California, 1962.

Polya, G., *Heuristic Reasoning in the Theory of Numbers,* Amer. Math. Monthly **66** (1959), 375–384.

11. The Prime Divisor Function, Selberg's Sieve Method, and Algebraic Independence

1. Introduction

In this chapter, we shall consider three different topics, the prime divisor function, Selberg s sieve method and algebraic independence. We shall first obtain the Euler product for $\sum_{n=1}^{\infty} 2^{\omega(n)}/n^s$, where $\omega(n)$ is the number of prime divisors of n. From this, it is easy to obtain a representation which allows us to derive the mean value. Then, using the fact that $\omega(n)$ is an additive function, we shall obtain its mean value. Then we obtain the mean value of the square of $\omega(n)$. With the aid of these results we can get a meaning to the statement about the average number of prime divisors of an integer.

Next, we turn to the sieve method of Selberg, giving his original proof. This is one of the most elegant and powerful methods in analytic number theory.

Finally, we give some results of algebraic independence. What we show is that the elementary arithmetic functions are algebraically independent. This requires some deep results in algebraic geometry.

2. The Prime Divisor Function

Let us now study the function defined by

$$(1) \qquad \omega(n) = \sum_{p|n} 1.$$

In words, $\omega(n)$ is the number of distinct prime divisors of n. This function is

Richard Bellman, Analytic Number Theory: An Introduction   ISBN 0-8053-0452-5

obviously additive. Consequently, we have the equation

(2)
$$\sum_{n=1}^{\infty} \frac{2^{\omega(n)}}{n^s} = \prod_p \left(1 + \frac{2}{p^s} + \cdots\right)$$

$$= \prod_p \left(\frac{1 + 1/p^s}{1 - 1/p^s}\right)$$

$$= \zeta^2(s)/\zeta(2s).$$

From this, we obtain the representation

(3)
$$2^{\omega(n)} = \sum_{k^2 \mid n} \mu(k) d(n/k^2).$$

Exercises

1. What do we get if we write $\zeta^2(s)/\zeta(2s)$ as $\zeta(s)(\zeta(s)/\zeta(2s))$?

2. Show that $2^{\omega(n)} \leqslant d(n)$, with equality only if n is squarefree.

3. The Mean Value of $2^{\omega(n)}$

Using the representation above we readily find

(1)
$$\sum_{n=1}^{N} 2^{\omega(n)} = c_1 N + 0(\sqrt{N}).$$

We leave it to the reader to determine the constant c_1.

4. The Mean Value of $\omega(n)$

We have

$$\sum_{n=1}^{N} \omega(n) = \sum_{n=1}^{N} \left(\sum_{p \mid n} 1\right)$$

(1)
$$= \sum_{p \leqslant n} \left[\frac{N}{p}\right]$$

$$= N \sum_{p \leqslant N} \frac{1}{p} + 0(N).$$

Using the prime number theorem, we have

(2)
$$\sum_{p \leqslant N} \frac{1}{p} = \log \log N + c_1 + 0(1).$$

Hence, we have the result

(3)
$$\sum_{n=1}^{N} \omega(n) = N \log \log N + 0(N).$$

However, we do not need the prime number theorem. An elementary argument suffices. We leave it as an exercise for the reader to supply it.

Exercises

1. Find the asymptotic order of $\Sigma_{n=1}^{N} \log d(n)$.

2. Find the asymptotic order of $\Sigma_{n=1}^{N} \log \phi(n)$.

3. Show that

$$\sum_{n=1}^{\infty} \frac{\omega(n)}{n^s} = \zeta(s) \sum_{p} \frac{1}{p^s}.$$

4. Find the Euler product for $\alpha^{\omega(n)}/n^s$.

5. By differentiation, obtain

$$\sum_{n=1}^{\infty} \frac{\omega(n)}{n^s} \quad \text{and} \quad \sum_{n=1}^{\infty} \frac{\omega^2(n)}{n^s}.$$

5. The Order of Magnitude of $\Sigma_{n=1}^{N} \omega^2(n)$

To evaluate this sum, we write

(1)
$$\sum_{n=1}^{N} \omega^2(n) = \sum_{n=1}^{N} \omega(n) \left(\sum_{p|n} 1 \right).$$

Inverting the order of summation and using the result of the previous exercises, we have

(2)
$$\sum_{n=1}^{N} \omega^2(n) = N(\log \log N)^2 + 0(N \log \log N).$$

Combining these results we have

(3)
$$\sum_{n=1}^{N} (\omega(n) - \log \log N)^2 = 0(N \log \log N).$$

From this result, we see a meaning we can attach to the statement, "Most numbers in the interval $(1, N)$ have $\log \log N$ prime divisors."

6. The Average Order of $\omega(g(n))$

Let us now consider the order of magnitude $\sum_{n=1}^{N} \omega(g(n))$ where $g(n)$ is an irreducible polynomial in n. We have

(1)
$$\sum_{n=1}^{N} \omega(g(n)) = \sum_{n=1}^{N} \left(\sum_{p \mid g(n)} 1 \right).$$

To estimate this last sum, we break the second sum into two parts, $1 \leqslant p \leqslant N$ and $p > N$. It is clear, since $g(n)$ is a polynomial that it has only a finite number of prime factors $p > N$. Consequently, we have

(2)
$$\sum_{n=1}^{N} \omega(g(n)) = \sum_{\substack{p \leqslant N \\ }} \left(\sum_{\substack{g(n) \equiv 0(p) \\ 1 \leqslant n \leqslant N}} 1 \right) + 0(N)$$

$$= N \sum_{p \leqslant N} \frac{\rho(p)}{p} + 0(N),$$

where, as before, $\rho(p)$ is the number of solutions of the congruence.

Thus, finally, we obtain the results

$$(3) \qquad \sum_{n=1}^{N} \omega(g(n)) = N \log \log N + 0(N).$$

Exercise

1. What do we get if the polynomial is reducible?

7. The Sieve Method of Selberg

Let us consider the following method of Selberg, which in many parts of number theory replaces the sieve method of Brun.

Suppose that we have a finite sequence of integers $\{a_k\}, k = 1, 2, \cdots, N$, and we wish to determine an upper bound for the number of the a_k which is not divisible by any prime $p \leqslant z$.

Let $\{x_\nu\}, 1 \leqslant \nu \leqslant z$, be a sequence of real numbers such that $x_1 = 1$, while the other x_ν are arbitrary. Consider the quadratic form

$$(1) \qquad Q(x) = \sum_{k=1}^{N} \left(\sum_{\nu | a_k} x_\nu \right)^2 .$$

Whenever a_k is not divisible by a prime $\leqslant z$, the sum $\sum_{\nu | a_k} x_\nu$ yields $x_1 = 1$. Hence, if we denote the number of a_k, $k = 1, 2, \cdots, N$, divisible by any prime $p \leqslant z$ by $f(N, z)$ we have

$$(2) \qquad f(N, z) \leqslant \sum_{k=1}^{N} \left(\sum_{\nu | a_k} x_\nu \right)^2$$

for all x_ν with $x_1 = 1$. Hence

$$(3) \qquad f(N, z) \leqslant \min_{x_\nu} \sum_{k=1}^{N} \left(\sum_{\nu | a_k} x_\nu \right)^2 .$$

To evaluate the right-hand side, we write

$$(4) \qquad Q(x) = \sum_{\nu_1, \nu_2 \leqslant z} x_{\nu_1} x_{\nu_2} \left\{ \sum_{\frac{\nu_1 \nu_2}{K} | a_k}^{1} \right\}$$

where a_k runs over the sequence $a_1, a_2, \cdots, a_N$. Here K denotes the greatest common divisor of v_1 and v_2.

Now suppose that the sequence $\{a_k\}$ possesses sufficient regularity properties so that there exists a formula of the type

(5)
$$\sum_{\substack{\rho \mid a_k \\ k=1,2,\cdots,N}} 1 = \frac{N}{f(\rho)} + R(\rho)$$

where $R(N)$ is a remainder term and the "density function" $f(\rho)$ is multiplicative, i.e.,

(6)
$$f(\rho_1\rho_2) = f(\rho_1)f(\rho_2) \quad \text{for } (\rho_1, \rho_2) = 1.$$

Then

(7)
$$\sum_{\frac{v_1 v_2}{K} \mid a_k} 1 = \frac{N}{f(v_1 v_2/K)} = \frac{f(K)N}{f(v_1)f(v_2)} + R\left(\frac{v_1 v_2}{K}\right).$$

Using this formula in (4), the result is

(8)
$$f(N, z) \leqslant \sum_{v_1, v_2 \leqslant z} \frac{x_{v_1} x_{v_2} f(K)}{f(v_1)f(v_2)} + \sum_{v_1, v_2 \leqslant z} x_{v_1} x_{v_2} R\left(\frac{v_1 v_2}{K}\right).$$

Consider now the problem of determining the x_v, $2 \leqslant v \leqslant z$, for which the quadratic form

(9)
$$P(x) = \sum_{v_1, v_2 \leqslant z} \frac{x_{v_1} x_{v_2} f(K)}{f(v_1)f(v_2)}$$

is a minimum.

To do this, we introduce the function

(10)
$$f_1(n) = \sum_{d \mid n} \mu(d)f(n/d).$$

If n is squarefree, we have

(11)
$$f_1(n) = f(n) \prod_{p|n} \left(1 - \frac{1}{f(p)}\right).$$

Using the Möbius inversion formula, (10) yields

(12)
$$f(K) = \sum_{n|K} f_1(n) = \sideset{}{'}\sum_{\substack{n|v_1 \\ n|v_2}} f_1(n).$$

Using this expression for $f(K)$ in (9), we see that

(13)
$$P(x) = \sum_{n \leq z} f_1(n) \left\{ \sum_{\substack{n|v \\ v \leq z}} \frac{x_v}{f(v)} \right\}^2.$$

Now perform a change of variable. Set

(14)
$$y_n = \sum_{\substack{n|v \\ v \leq z}} \frac{x_v}{f(v)}.$$

Then, using the Möbius inversion formula once more, we have

(15)
$$\frac{x_v}{f(v)} = \sum_{n \leq z/v} \mu(n) y_{nv}.$$

The problem is thus that of minimizing

(16)
$$P = \sum_{n < z} f_1(n) y_n^2$$

over all $\{y_k\}$ such that

(17)
$$\sum_{n < z} \mu(n) y_n = \frac{\lambda_1}{f(1)} = 1.$$

It is easily seen that the minimizing $\{y_k\}$ is given by

(18)
$$y_n = \frac{\mu(n)}{f_1(n)} \frac{1}{\sum\limits_{\rho' \leqslant z} \frac{\mu^2(\rho')}{f_1(\rho')}}$$

and that the minimum value of the form P is

(19)
$$\frac{1}{\sum\limits_{n \leqslant z} \frac{\mu^2(n)}{f_1(n)}} .$$

The corresponding x_n are determined by the relations

(20)
$$x_n = \frac{f(n)}{\sum\limits_{\rho \leqslant z} \frac{\mu^2(\rho)}{f_1(\rho)}} \sum\limits_{\rho \leqslant z/n} \frac{\mu(\rho)\mu(\rho n)}{f_1(\rho n)}$$

$$= \mu(n) \prod\limits_{p|n}\left(1 - \frac{1}{f(p)}\right)^{-1} \frac{1}{\sum\limits_{\rho \leqslant z} \frac{\mu^2(\rho)}{f_1(\rho)}} \sum\limits_{\substack{\rho \leqslant z/n \\ (\rho,n)=1}} \frac{\mu^2(\rho)}{f_1(\rho)} .$$

Inserting these values in (8), we obtain the inequality

(21)
$$f(N, z) \leqslant \frac{N}{\sum\limits_{\rho \leqslant z} \frac{\mu^2(\rho)}{f_1(\rho)}} + \sum\limits_{\nu_1, \nu_2 \leqslant z} |x_{\nu_1} x_{\nu_2} R(\nu_1 \nu_2/K)| .$$

If z is chosen properly, say $N^{1/2 - \epsilon}$, the upper bound is nontrivial.

A particularly simple case is that where $a_k \equiv k$, so that $f(\rho) \equiv \rho$. We then obtain a bound for the number of primes less than or equal to N.

Exercises

1. Prove that the number of integers less than x which have exactly k prime factors is bounded by

$$\frac{c_1 x (\log \log x + c_2)^{k-1}}{(k-1)! \log x}$$

2. Prove that the number of integers n such that n and $n + 2$ are both prime, twin primes, less than x is bounded by $c_3 x/(\log x)^2$.

3. Hence, prove that $\Sigma\, 1/n$, where the summation is over twin primes, converges.

4. Consider the Goldbach representation, $p_1 + p_2 = n$, where p_1 and p_2 are primes. Prove that the number of such representations is bounded by $c_4 n/(\log n)^2$.

5. Obtain a bound for the number of primes in the arithmetic progression $an + b$.

6. Obtain the bound for the number of primes in the arithmetic progression $an + b$ in the interval $(x, x + y)$.

7. Find the order of magnitude of $\Sigma_{p \leqslant n}\, d(p + 2)$.

8. Extend Selberg's method of the case where the quadratic form is of the type $\Sigma_{n=1}^{N} f(n)(\cdots)^2$.

8. The Algebraic Independence of Arithmetic Functions

The search for relationships between various arithmetic functions has, in the past, resulted in many interesting identities. These efforts have been centered on the problem of evaluating various types of sums of arithmetic functions in terms of others. In this section we will adopt a negative approach to the problem, and consider the question of when one can assert the algebraic independence of a set of arithmetic functions.

Before proceeding to discuss what types of arithmetic functions we intend to consider, we shall make these terms more precise.

DEFINITION 1. A single-valued function $f(n)$ will be called arithmetic if it is defined to have real values for all positive integral values of n.

DEFINITION 2. An arithmetic function $f(n) \neq 0$ will be called multiplicative if, for m and n relatively prime, $f(mn) = f(m)f(n)$.

Perhaps the best known examples of multiplicative arithmetic functions are n; $d(n)$, the number of divisors of n; $\sigma(n)$, the sum of the divisors of n; $\phi(n)$, the Euler function; $\mu(n)$, the Möbius function; $2^{\omega(n)}$ where $\omega(n)$ is the number of distinct primes dividing n.

DEFINITION 3. A set of arithmetic functions $f_1, \cdots, f_N$ will be called algebraically independent over the real field if there exists no polynomial $P(x_1, \cdots, x_N) \neq 0$ with real coefficients, irreducible over the real field, such that

$$P(f_1, \cdots, f_N) = 0,$$

for all positive integral values of n.

In this section we discuss only multiplicative arithmetic functions.

An interesting product of the investigations of this section is the following corollary:

The functions n, $\phi(n)$, d(n), $2^{\omega(n)}$, $\sigma(n)$ *and* $\mu(n)$ *are algebraically independent.*

The proof of this result is circuitous in that it seems essential to discuss algebraic relationships between general multiplicative arithmetic functions first.

Since the polynomial relations between two multiplicative functions can be discussed more fully than those involving more than two functions, this case will be discussed first. The extension of some of the results to more than two functions depends upon the results obtained for the case of two.

To begin with, it should be noted that any two arithmetic functions $f(n)$, $g(n)$, each of which takes on only a finite number of different values are algebraically dependent. This follows from the fact that there are only a finite number of points $[f(n), g(n)]$ in the plane, and by Bertini's theorem an irreducible algebraic curve can be passed through the points.

DEFINITION 4. A function $f(n)$ which assumes only a finite number ot differerent values for all positive integral n, will be said to be finitely valued.

DEFINITION 5. A function $f(n)$ which assumes an infinite number of distinct values on a set of numbers of the form p^{a_k}, for only a finite number of primes p, will be called singular.

DEFINITION 6. A set of functions $f_k(n)$, $2 \leqslant k \leqslant N$, will be said to be singular if all the $f_k(n)$ simultaneously take on an infinite number of distinct values only on a set p^{a_k}, where p is a fixed prime for all k.

Also to be noted at this point is the fact that if two arithmetic functions $f(n)$ and $g(n)$ constitute a singular set, they may be algebraically dependent, even under the added condition that f and g be multiplicative. To show this we make use of the fact that a multiplicative function may be defined arbitrarily at the values $n = p^k$, $k = 1, 2, \cdots, p$ a prime. Consider any irreducible algebraic curve through the points $[0, 0]$ and $[1, 1]$ and for some fixed prime p let the couples $[f(p^{\alpha_i}), g(p^{\alpha_i})]$ correspond to an infinite set of points on the curve. For all other $n \neq 1$ define $f(n)$ and $g(n)$ to be zero. These functions are then clearly both multiplicative and algebraically dependent.

We can moreover give an example of three multiplicative arithmetic functions, forming a singular set, which are positive, and still algebraically dependent. Suppose that $[a_i, b_i, c_i]$ are the finite number of sets of values assumed by $[f, g, h]$ for the values $n \neq p^{\alpha_i}$, $(n, p) = 1$. Then let all the points $[f(p^{\alpha_i}), g(p^{\alpha_i}), h(p^{\alpha_i})]$ lie on an algebraic curve C through the points $[a_i, b_i, c_i]$ and consider the linear transformations

$$t_i: \begin{bmatrix} a_i & 0 & 0 \\ 0 & b_i & 0 \\ 0 & 0 & c_i \end{bmatrix}.$$

Operating with the inverses of these we get a set of curves $C_i = t_i^{-1}(C)$. Then by Bertini's theorem we can pass an irreducible algebraic surface S through the curves C and C_i. The surfaces $t_i(S)$ will all intersect in the curve C. Since the points $[f, g, h]$ lie on S, f, g, and h are algebraically dependent, and the above construction insures complete consistency with the multiplicativity of these functions.

The above remarks indicate the nature of some of those cases of multiplicative functions which are algebraically dependent. They also serve to establish somewhat the necessity of the hypotheses which will be found in the following theorems.

We will now state some auxiliary results which will be needed later.

LEMMA 1. *If* f(n) *is not finitely valued there is no polynomial relation* P[f(n)] = 0.

This is clear.

LEMMA 2. *If* f(n) *is a multiplicative arithmetic function, then either (a)* f(n) *is finitely valued, or (b)* f(n) *is singular, or (c) there exists a sequence of integers* δ_k *relatively prime by pairs such that* $f(\delta_k)$ *takes on an infinite number of distinct values.*

The proof of this follows easily by showing that the negation of (a) and (c) implies (b).

THEOREM 1. *If* f(n) *and* g(n) *are two multiplicative arithmetic functions such that* [f, g] *is not singular, and at least one of the two is not finitely valued, then* f *and* g *are algebraically independent, unless* $f = g^r$ *where* r *is a rational number.*

PROOF. Suppose that f and g are algebraically dependent, so that $P[f, g] = 0$ where $P(x, y)$ is an irreducible polynomial.

If $f(n)$ is finitely valued, then $g(n)$ takes on an infinite number of different values on a set n_i, $i = 1, \cdots$. We write

$$P[f, g] = P_m [f] g^m + \cdots + P_0 [f].$$

Then since f is finitely valued there is an infinite subsequence of the n_i such that $f'(n_i) = A$, a constant, and $g'(n_i)$ assumes an infinite number of different values on this subsequence. But then

$$P_m [A] g^m + \cdots + P_0 [A] = 0$$

would have an infinite number of roots, which in turn implies that

$$P_k [A] = 0, \quad (k = 0, \cdots, m).$$

Hence

$$P_k(x) = (x - A)Q_k(x)$$

and

$$P(x, y) = (x - A)Q(x, y),$$

contrary to the assumption that $P(x, y)$ is irreducible.

On the other hand if both f and g are not finitely valued, Lemma 2 implies that there exists either (a) two infinite sequences $\{\alpha_i\}$, $\{\beta_i\}$ such that $(\alpha_i, \beta_i) = 1$ for all i and j, and at least one of the two functions, say g, takes on an infinite number of distinct values on *both* $\{\alpha_i\}$ and $\{\beta_i\}$; or (b) two primes p and q such that $p \neq q$, and f takes on an infinite number of distinct values on a sequence p^{α_i}, and g takes on an infinite number of distinct values on a sequence q^{β_i}.

Now we have $P[f(n), g(n)] = 0$, and if $(m, n) = 1$, $f(m) = a$, $g(m) = b$,

$$P[af(n), bg(n)] = 0.$$

Case (a). We fix the couple (a, b) as any element of the sequence $[f(\alpha_i), g(\alpha_i)]$ and let n run through the sequence $\{\beta_i\}$. Then $P[f, g] = 0$ and $P[af, bg] = 0$ have an infinite number of roots in common. Eliminate f between these two equations thus obtaining the resultant $R(g)$, which must vanish for the infinity of values $g(\beta_i)$. This clearly implies that $R(g) = 0$, since R is a polynomial in g. Since $P[f, g]$ is irreducible we must have

$$P[af, bg] = c(a, b)P[f, g],$$

or

$$\sum A_{ij}a^i b^j f^i g^j = c(a, b) \sum A_{ij} f^i g^j,$$

and thus $a^i b^j = c(a, b)$. Since $P[f, g]$ is irreducible and must contain both variables it must have more than one term. Hence, we get an equation of the form $a^i b^j = a^k b^t$ which we can write as

$$a^r b^s - 1 = 0 \quad \text{or} \quad a^r - b^s = 0.$$

Thus in either case $Q(x, y) = x^r y^s - 1$, or $x^r - y^s$, has a degree less than or equal to that of $P(x, y)$. Now since (a, b) can run through the infinity of couples $[f(\alpha_i), g(\alpha_i)]$, $Q(x, y)$ and $P(x, y)$ have an infinite number of roots in common. Eliminating x, we again get that the resultant $R_1(y) = 0$. But this implies that $P(x, y)$ divides $Q(x, y)$, and since the degree of $Q(x, y)$ is less than or equal to the degree of $P(x, y)$ we must have $P(x, y) = cQ(x, y)$, and hence $Q(f, g) = 0$ which is the conclusion of the theorem.

Case (b). The argument is much the same as in Case (a). (a, b) is first fixed in the sequence $[f(p^{\alpha_i}), g(p^{\alpha_i})]$ and we let n run through the q^{β_i}. Then as before we get $Q(x, y)$ to have the same forms as noted above, having the roots $[f(p^{\alpha_i}), g(p^{\alpha_i})]$ in common with $P(x, y)$. Then proceeding from this point we now eliminate y between $Q(x, y)$ and $P(x, y)$ obtaining the resultant $R_2(x)$. But since $R_2[f(p^{\alpha_i})] = 0$, and $f(p^{\alpha_i})$ takes on infinitely many values, $R_2(x) = 0$, and the conclusion follows as before.

COROLLARY. *If* f(n) *is a multiplicative arithmetic function, and* n *and* f(n) *are algebraically dependent, then* f(n) = n^r *for some fixed rational number* r.

Before proceeding to the general case of any number of multiplicative arithmetic functions, there is one more pathological case to mention produced by our definitions of r. For $N > 2$, we can have a set of arithmetic functions such as the following:

$$f_1(p_1^\alpha) = c_{\alpha_1} \neq 0, \quad f_1(1) = 1, \quad f_1(n) = 0 \quad \text{otherwise;}$$

$$f_2(p_2^\alpha) = c_{\alpha_2} \neq 0, \quad f_2(1) = 1, \quad f_2(n) = 0 \quad \text{otherwise;}$$

$$f_3(p_3^\alpha) = c_{\alpha_3} \neq 0, \quad f_3(1) = 1, \quad f_3(n) = 0 \quad \text{otherwise.}$$

If the p_1, p_2, p_3 are different primes and the c_{α_i} are any infinite sequence of real numbers, f_1, f_2 and f_3 clearly constitute a nonsingular set of multiplicative arithmetic functions. Also, they are algebraically dependent since

$$Q[f_1, f_2, f_3] = f_1 f_2 + f_2 f_3 - 2 f_1 f_3,$$

is an irreducible form, and vanishes for all n.

For obvious reasons our hypotheses will have to be formulated so as to eliminate such sets of functions from consideration. Due to the variety of peculiar combinations of functions which can arise, as is indicated by the above discussion, we will not be able to prove as sharp a result for the case $N > 2$ as in Theorem 1. However, we can obtain a workable result.

THEOREM 2. *Let* $f_k(n)$, k = 1, 2, $\cdots$, N, *be a set of multiplicative arithmetic functions. If a polynomial* $P(x_1, \cdots, x_N) \neq 0$, *reducible or irreducible, vanishes on the sets* $(f_1(n), \cdots, f_N(n))$, n = 1, 2, $\cdots$, *then for every sequence of pairwise relatively prime integers* $\{n_i\}$ *there exists a subsequence* $\{n_i'\}$ *on which there holds a relation of the form*

(1) $$f_1^{\alpha_1} \cdots f_N^{\alpha_N} = f_1^{\beta_1} \cdots f_N^{\beta_N},$$

where the α_i, β_i *are nonnegative integers and, for at least one* i, $\alpha_i \neq \beta_i$.
In (1) it is understood that for k = 0, $f^k = 1$. *Also a relation*

(2) $$f_1^{\alpha_1} \cdots f_N^{\alpha_N} = 0$$

on $\{n_i'\}$ is of the form (1), i.e., $f_1^{\alpha_1} \cdots f_N^{\alpha_N} = f_1^{2\alpha_i} \cdots f_N^{2\alpha_N}$ on $\{n_i'\}$.

PROOF. Suppose that $P(f_1(n), \cdots, f_N(n)) = 0$ for all n. Let $\{n_i\}$ be any sequence of relatively prime integers. If P has only one term, the theorem is proved by the note above. If P has more than one term, write $P(f_1(n), \cdots, f_N(n)) = Af_1^{\alpha_1}(n) \cdots f_N^{\alpha_N} + Bf_1^{\beta_1} \cdots f_N^{\beta_N} + \cdots$. Either (1) holds on $\{n_i\}$ or there is an $m \in \{n_i\}$ such that $a = f_1^{\alpha_1}(m) \cdots f_N^{\alpha_N}(m) \neq f_1^{\beta_1}(m) \cdots f_N^{\beta_N}(m) = b$. In the second case, $P(f_1(mn), \cdots, f_N(mn)) = Aaf_1^{\alpha_1}(n) \cdots f_N^{\alpha_N}(n) + Bbf_1^{\beta_1}(n) \cdots f_N^{\beta_N}(n) + \cdots$ and $aP(f_1(n), \cdots, f_N(n)) - P(f_1(mn), \cdots, f_N(mn)) = B(a - b)$ $f_1^{\beta_1}(n) \cdots f_N^{\beta_N}(n) + \cdots$ is not identically zero, $a \neq b$, vanishes for all n such that $(n, m) = 1$, and has fewer terms than P. It is clear that either (1) holds on some $\{n_i'\}$ or the process can be continued until the resulting polynomial is of the form (2) and vanishes for all n with $(n, m_i) = 1$, $i = 1, \cdots, k$, $k <$ the number of terms of P. This completes the proof.

We will now apply the above results to certain rather well-known sets of arithmetic functions.

THEOREM 3. *If $f_{k+1} = \Sigma_{d|n}f_k(d)$, $k = 0, \cdots, N$, $f_0(n)$ is multiplicative and $f_0(p)$ takes on an infinite set of values over some set of primes, then $f_0, \cdots, f_N$ are algebraically independent.*

PROOF. Since $f_0(n)$ is multiplicative, all the $f_k(n)$ are multiplicative. Also since $f_0(p)$ takes on an infinite number of distinct values on the sequence of primes, so does $f_i(p) = f_{i-1}(p) + 1 = f_0(p) + i$. Thus if $f_0, \cdots, f_N$ were to be algebraically related we would have

$$[f_0(p) + i_1]^{\alpha_1} \cdots [f_0(p) + i_k]^{\alpha_k} = [f_0(p) + i_{k+1}]^{\alpha_{k+1}} \cdots [f_0(p) + i_R]^{\alpha_R},$$

where $i_j \neq i_j$, $0 \leq i_j \leq N$, $\alpha_{ij} > 0$. But since $f_0(p)$ takes on an infinite number of distinct values this would require that we have

(3) $$\{x + i_1]^{\alpha_1} \cdots [x + i_k]^{\alpha_k} = [x + i_{k+1}]^{\alpha_{k+1}} \cdots [x + 1_R]^{\alpha_R}.$$

But this is impossible, for one side has the root $x = -i_1$, and the other side does not. Thus by Theorem 2, $f_0, \cdots, f_N$ are algebraically independent.

COROLLARY. n, $\phi(n)$, $\sigma(n)$ *are algebraically independent.*

PROOF. Take $f_0 = \phi(n)$ in the theorem. Then $f_1 = n$, $f_2 = \sigma(n)$.

THEOREM 4. n, $\phi(n)$, $\sigma(n)$, d(n), $2^{\omega(n)}$ *are algebraically independent.*

PROOF. We first show that there exists an infinite sequence of primes satisfying the following five conditions:

(1) $p_i \equiv -1 \pmod 3$, $p_i > 3$;

(2) $p_i \not\equiv 1 \pmod{p_j}$;

(3) $p_i \neq 2^\alpha + 1$;

(4) $p_i^2 + p_i + 1 \not\equiv 0 \pmod{p_j}$ and mod 3);

(5) $p_j \equiv 1 \pmod 2$, $p_j - 1 \equiv 0 \pmod{p_j}$ implies that $p_i^2 + p_i + 1 \not\equiv 0 \pmod{p_j}$.

We begin by choosing p_1 as any prime of the form $3n - 1$ which is not equal to $2^\alpha + 1$. Suppose then that $p_1, \cdots, p_r$ have already been constructed. We will now establish the existence of a suitable p_{r+1}. Let 2^{δ_i} be the highest power of 2 dividing $p_i - 1$, and set

$$\gamma = 3 \prod_{r=1}^{r} (p_i^2 + p_i + 1) p_i \{(p_i - 1)/2^{\delta_i}\}.$$

If we consider the form $\gamma x - 1$, Dirichlet's theorem gives us a prime of this form, which is greater than all the p_i and all the prime divisors of $p_i^2 + p_i + 1$, $i = 1, \cdots, r$. We let this prime be p_{r+1}. Clearly p_{r+1} satisfies (1). That it can be chosen to satisfy (3) also follows easily from the fact that $\Sigma p^{-1} \cdot \log p$ ($p \equiv -1 \pmod \gamma$) diverges whereas $\Sigma p^{-1} \log p$ ($p = 2^\alpha + 1$) converges. (2) follows since

$$p_{r+1} - 1 = \gamma x - 2 \equiv -2 \not\equiv 0 \pmod{p_i},$$

where $i = 1, \cdots, r$. It is obvious that $p_i^2 + p_i + 1 \not\equiv 0 \pmod{p_{r+1}}$ and $p_{r+1}^2 + p_{r+1} + 1 \equiv 1 \not\equiv 0 \pmod 3$ so that (4) is satisfied. Finally if s_j is any odd divisor of $p_j - 1$, $j \leqslant N$,

$$p_{r+1} \equiv -1 \pmod{s_1}$$

so that $p_{r+1} - 1 \equiv -2 \not\equiv 0 \pmod{s_j}$.

$$p_{r+1}^2 + p_{r+1} + 1 \equiv 1 \not\equiv 0 \pmod{s_j}.$$

In addition if $s_{r+1} \equiv 1 \pmod 2$, $s_{r+1} | p_{r+1} - 1$ and $s_{r+1} | p_i^2 + p_i + 1$, these imply $s_{r+1} | \gamma$ and hence $p_{r+1} \equiv -1 \equiv 1 \pmod{s_{r+1}}$, which is impossible. This gives us the desired infinite sequence of primes $p_1, \cdots, p_r, p_{r+1}, \cdots$.

We now form from these primes the numbers

$$\delta_1^* = p_1^2, \quad \delta_2^* = p_2^2 p_3^2, \cdots, \quad \delta_k^* = p_{r+1}^2 \cdots p_{r+k}^2, \quad r = \frac{k(k-1)}{2}.$$

Since the functions simultaneously take on an infinite number of distinct values on the sequence $\{\delta_k^*\}$, and since the δ_i^* are relatively prime to one another, we are now in a position to apply Theorem 2. Suppose then that

$$n^{\alpha_1} (\phi(n))^{\alpha_2} (\sigma(n))^{\alpha_3} (d(nn))^{\alpha_4} 2^{\omega(n)\alpha_5} = n^{\beta_1} (\phi(n))^{\beta_2} (\sigma(n))^{\beta_3} (d(nn))^{\beta_4} 2^{\omega(n)\beta_5}$$

over some subsequence of the $\{\delta_k\}$. This gives

$$(\Pi p^2)^{\alpha_1}(\Pi p(p-1))^{\alpha_2}(\Pi(p^2+p+1))^{\alpha_3}3^{k\alpha_4}2^{k\alpha_5}$$
$$= (\Pi p^2)^{\beta_1}(\Pi p(p-1))^{\beta_2}(\Pi(p^2+p+1))^{\beta_3}3^{k\beta_4}2^{k\beta_5}.$$

From the manner in which the p's were chosen, each $p_i - 1$ contains an odd prime divisor greater than 3 which divides none of the other terms which appear. Hence $\alpha_2 = \beta_2$. Then since $p_i > 3$, and divides no term of the form $p_i^2 + p_i + 1$, we must have $\alpha_1 = \beta_1$. Similarly since $p_i^2 + p_i + 1 \not\equiv 0 \pmod 2$, $\pmod 3$, we get $\alpha_3 = \beta_3$. This leaves $3^{\alpha_4}2^{\alpha_5} = 3^{\beta_4}2^{\beta_5}$ so that $\alpha_4 = \beta_4$ and $\alpha_5 = \beta_5$. Thus by Theorem 2, the set of functions under consideration are algebraically independent.

COROLLARY. *The set of functions* n, $\phi(n)$, $\sigma(n)$, $d(n)$, $2^{\omega(n)}$, $\mu(n)$, *are algebraically independent.*

[$\mu(n)$ is the Möbius function. Since it has a range of only the three values 0, -1, 1, its algebraic independence of other sets of functions cannot be obtained directly from Theorem 2.]

PROOF. Let $P(f_1, \cdots, f_6) = 0$ be an irreducible polynomial relationship between the functions mentioned in the corollary. If we let $f_6 = \mu(n)$ we may write

$$P = [\mu(n)]^r P_r[f_1, \cdots, f_5 + \cdots + \mu(n)P_1[f_1, \cdots, f_5] + P_0[f_1, \cdots, f_5] = 0.$$

But over the sequence $\{\delta_k^*\}$ constructed in the proof of the theorem $\mu(\delta_k^*) = 0$. Thus over the sequence $\{\delta_k^*\}$ constructed in the proof of the theorem we must have $P_0[f_1, \cdots, f_5] = 0$. But this in turn is impossible by Theorem 4. Hence $P_0[f_1, \cdots, f_5] = 0$ which is a contradiction since then $P[f_1, \cdots, f_6]$ is divisible by f_6 and hence not irreducible.

Exercise

1. Prove that the assumption that $f(n)$ is multiplicative is unnecessary.

Bibliography and Comments

Section 1

For further information, see the book

Hardy, G. H., and E. M. Wright, *Introduction to the Theory of Numbers,* Clarendon Press, Oxford, 5th ed. 1979.

See

Dressler, R. E., and J. van de Lune, *Some Remarks Concerning the Number Theoretic Functions* $\omega(n)$ *and* $\Omega(n)$, Proc. Amer. Math. Soc. **41** (1973), 403–406.
Mirsky, L., *On the Distribution of Integers having a Prescribed Number of Divisors,* Simon Stevin **26** (1949), 168–175.
Tijdeman, R., *Old and New Number Theory,* Nieuw Arch. Wisk **20** (3) (1972), 20–30.

Section 2

See

Chowla, S. D., and T. Vijayaraghavan, *On the Largest Prime Divisors of Numbers,* J. Indian Math. Soc. (N.S.) **11** (1947), 31–37.

Section 3

See

Grosswald, E., *The Average Order of an Arithmetic Function,* Duke Math. J. **23** (1956), 41–44.
Yüh, Ming-I., *A Note on an Arithmetical Function,* Sci. Record (N.S.) **1** (2) (1957), 9–12.

Section 5

We are using methods from probability theory, particularly the well-known inequality of Cebycev. In general, the methods of one theory can be used in the other. See

Renyi, A., *Methods of Probability Theory in Number Theory,* McGraw-Hill 1955.

The result concerning average order is due to Hardy and Ramanujan. The proof given here is due to Turan.
For the concept of normal order of an arithmetic function, see the book by Hardy and Wright.

Section 7

The original paper of Selberg is

Selberg, A., *On an Elementary Method in the Theory of Primes*, Norske Vid. Selsk. Forh. (Trondheim) **19** (18) (1947).

A more accessible source is

Bellman, R., *Introduction to Matrix Analysis*, McGraw-Hill, New York, 1960; 2nd ed., 1970.

See also

Bellman, R., *Dynamic Programming and the Quadratic Form of Selberg*, J. Math. Anal. Appl. **15** (1966), 30–32.

Birman, A., *On the Existence of a Co-relation Between Two Unsolved Problems in Number Theory*, Riveon Lematematika **13** (1959), 17–19.

Section 8

These results were presented in

Bellman, R., and H. N. Shapiro, *The Algebraic Independence of Arithmetic Functions*. I, *Multiplicative Functions*, Duke Math. J. **15** (1948), 229–235.

For further results see,

Carlitz, L., *Independence of Arithmetic Function*, Duke Math. J. **19** (1952), 65–70.

Popken, J., *Algebraic Dependence of Arithmetic Functions*, Nederl. Akad. Wetensch. Proc. Ser. A **65** – Indag. Math. **24** (1962), 155–168.

Wade, L. J., *Algebraic Independence of Certain Arithmetic Functions*, Duke Math. J. **15** (1948), 237.

As the paper by Popken indicates, we are entering the area pioneered by Liouville.

Ritt, J. F., *Integration in Finite Terms*, Columbia University Press, New York, 1948.

12. Tauberian Theorems

1. Introduction

In this chapter we wish to consider some Tauberian theorems. Tauberian theorems may be thought of as a method for converting from one average to another, and thus are an indispensable tool for the analytic number theorist.

In the first few sections, we give the fundamental results of Hardy and Littlewood. Then we apply the continuous version of this result to the error term for $\Sigma_{n=1}^{N} \phi(n)$ and the mean value of $\zeta^2(s)$ along the critical line.

In the final section we say a few words about the Erdös-Selberg method in the theory of primes.

2. The Tauberian Theorem of Hardy and Littlewood

It is elementary to go from the asymptotic order of $\Sigma_{n=1}^{N} a_n$ to the behavior of $\Sigma_{n=1}^{\infty} a_n x^n$ as $x \to 1$. A result of this type is called Abelian. To go in the reverse direction we require some condition on a_n. Hardy and Littlewood obtained many results of this type. Basically, we can pass from the behavior of $\Sigma_{n=1}^{\infty} a_n x^n$ as $x \to 1$ to the behavior of $\Sigma_{n=1}^{N} a_n$ if we know that $a_n \geqslant 0$.

A fundamental result of Hardy and Littlewood is that if

$$(1) \qquad \sum_{n=1}^{\infty} a_n x^n \sim \frac{1}{(1-x)^a} \left\{ \log\left(\frac{1}{1-x}\right) \right\}^b \quad \text{as } x \to 1$$

Richard Bellman, Analytic Number Theory: An Introduction   ISBN 0-8053-0452-5

and $a_n \geqslant 0$, then we have

(2)
$$\sum_{n=1}^{N} a_n \sim \frac{N^a (\log N)^b}{\Gamma(a+1)}, \quad a \geqslant 0.$$

Exercises

1. Prove that

$$(1-x) \sum_{n=0}^{\infty} a_n x^n = \sum_{n=0}^{\infty} (a_n - a_{n-1}) x^n, \quad a_{-1} = 0.$$

2. Hence show that we can use a Tauberian theorem if we add the hypothesis that a_n is monotone increasing.

3. Write

$$\frac{1}{(1-x)(1-x^2)\cdots(1-x^n)\cdots} = \sum_{n=0}^{\infty} p_n x^n.$$

What is the significance of the coefficient p_n?

4. Show that p_n is monotone increasing and use an appropriate Tauberian theorem to obtain the asymptotic behavior of p_n.

5. On the basis of the prime number theorem, find the behavior of $\Sigma_p \, x^p$ as $x \to 1$.

6. Show that

$$\sum_{p}(x^p)^2 = \sum_{n=1}^{\infty} b_n x^n,$$

where b_n is the Goldbach number, the number of solutions of $p_1 + p_2 = n$, where p_1 and p_2 are primes.

7. Find the order of magnitude of $\Sigma_{n=1}^{N} b_n$ as $N \to \infty$.

3. Continuous Version

The continuous version of the foregoing result is also useful. If we have

(1)
$$\int_{0}^{\infty} f(x) e^{-xy} dx \sim \frac{(\log y)^b}{y^a} \quad \text{as } y \to 0,$$

and $f(x) \geqslant 0$, then

(2)
$$\int_0^T f(x)\,dx \sim \frac{T^a (\log T)^b}{\Gamma(a+1)} \quad a > 0.$$

Exercises

1. Consider the equation $u(x) = f(x) + \int_0^x k(x-y)u(y)\,dy$, commonly called the renewal equation. Show that $u(x)$ is nonnegative if $f(x)$ and $k(x)$ are non-negative.

2. Show that this equation can be solved in terms of the Laplace transform of $u(x)$.

3. Obtain conditions on $f(x)$ and $k(x)$ which permit the determination of the asymptotic behavior of $\int_0^T u(x)\,dx$. (This equation plays a very important role in the theory of branching processes. See the book

Mode, C. J., *Multitype Branching Processes*, American Elsevier, New York, 1971.

Other information concerning the renewal equation can be found in

Bellman, R. and K. L. Cooke, *Differential-Difference Equations*, Academic Press, New York, 1963.

4. The Mean Value of $|\zeta(1/2 + it)|^2$

We begin with the fundamental representation

(1)
$$\Gamma(s)\zeta(s) = \int_0^\infty \frac{x^{s-1}\,dx}{e^x - 1}.$$

Regarding this as a Mellin transform, we have the inversion formula

(2)
$$\frac{1}{e^x - 1} = \frac{1}{2\pi i} \int_C x^{-s}\Gamma(s)\zeta(s)\,ds.$$

Here C is a straight line parallel to the imaginary axis to the right of the line $\sigma = 1$.

Let us now shift the contour. If we shift past the line $\sigma = 1$, we encounter a simple pole at $s = 1$. This result is the formula

(3)
$$\frac{1}{e^x - 1} - \frac{1}{x} = \frac{1}{2\pi i} \int_{C_1} x^{-s}\Gamma(s)\zeta(s)\,ds.$$

The contour C_1 is a line $\sigma = \frac{1}{2}$. It remains to choose x adroitly. Let us choose the value $x = ie^{i\sigma}$. If we use the fact that $i = e^{\pi i/2}$ we see that (3) is a Fourier transform.

We now use the Parseval-Plancheral formula. The details are given in a monograph by Titchmarsh cited at the end of the chapter.

5. The Error Term in $\Sigma_{n=1}^{N} d(n)$

Let $d(n)$ denote the number of divisors of n, and consider

$$(1) \qquad\qquad D(x) = \sum_{n<x} d(n).$$

It was proved by Dirichlet that

$$(2) \qquad D(x) = x \log x + (2C - 1)x + \Delta(x) \qquad (x \to \infty),$$

where $\Delta(x) = O(x^{1/2})$, and C is Euler's constant. Subsequently, it was shown by Voronoi that $\Delta(x) = O(x^{1/3} \log x)$, and the estimate has been continually improved since, although the precise result is still unknown.

In the other direction, it was proved by Hardy that for some constant k,

$$(3) \qquad\qquad |\Delta(x)| \geqslant k x^{1/4}$$

for an infinity of values of $x \to \infty$, and even that

$$(4) \qquad\qquad |\Delta(x)| \geqslant k(x \log x)^{1/4} \log\log x$$

for an infinity of $x \to \infty$. A result of this type is called an Ω-result, and (4) is written

$$(5) \qquad\qquad \Delta(x) = \Omega(x \log x)^{1/4} (\log\log x).$$

If we are interested, not in exceptional values, but in the average value, as in the case of $d(n)$ itself, we have the following result, also due to Hardy

$$(6) \qquad\qquad \int_1^T |\Delta(x)| dx = O(T^{5/4+\epsilon}) \qquad (T \to \infty).$$

The result we wish to prove in this chapter is a refinement of this result, namely

(7)
$$\int_1^T \frac{\Delta(x)^2}{x^{3/2}} dx \sim c_1 \log T \quad (T \to \infty),$$

where $c_1 \neq 0$ is a constant. Hence $\Delta(x) = \Omega(x^{1/4})$.

The same method as used in the proof of this theorem yields a similar result for the error term in the Gauss circle problem, which concerns itself with the asymptotic order of

(8)
$$\sum_{n<x} r(n) = R(x)$$

where $r(n)$ is the number of solutions in positive integers of $a^2 + b^2 = n$. We shall consider the error term in the estimate of

(9)
$$\sum_{n<x} \phi(n) = F(x),$$

where $\phi(n)$ is the Euler ϕ-function, and derive a result similar to that above.

We mention in passing that this method yields the known results for the functions $\pi(x) - Li(x)$ and $\psi(x) - x$ as far as Ω-results are concerned, but not the mean-value theorems, which seem to require more knowledge concerning the distribution of the zeros of the Riemann ζ-function (see

Ingham, A. E., *The Distribution of Prime Numbers,* Cambridge University Press, London, 1932.

The method used in this section for dealing with problems of this type seems due to Titchmarsh.

We have

(10)
$$\zeta^2(s) = \sum_{n=1}^{\infty} \frac{d(n)}{n^s} \quad (Re(s) > 1).$$

Using Perron's sum formula, we deduce

(11)
$$\sum_{n<x} d(n) = \frac{1}{2\pi i} \int_{2-i\infty}^{2+i\infty} \zeta^2(w) \frac{x^w}{w} dw \quad (x > 0, \text{nonintegral}).$$

Since the integer values of x form a denumerable set, we can ignore the behavior of the sum-function at these points.

We now shift the line of integration to the left of the line $2 + it$ until we reach the line $a + it$, $\frac{1}{4} < a < \frac{1}{2}$. This is accomplished by the standard method of first considering the rectangle whose vertices are $2 \pm iT$, $a \pm iT$, and then allowing T to tend to ∞. There is no difficulty in justifying this procedure using the order relation $|\zeta(\frac{1}{2} + it)| = O(|t|^{\frac{1}{4}})$ as $t \to \infty$. The only singularity of the integrand is at $w = 1$, and the residue at this point

$$(12) \qquad R(x) = x \log x + (2C - 1)x.$$

Thus

$$(13) \qquad \sum_{n < x} d(n) - R(x) = \frac{1}{2\pi i} \int_{-\infty}^{\infty} \frac{\zeta^2(a + it)}{(a + it)} x^{a+it} dt.$$

Let us now make the substitution $x = e^u$, so that the right side is a Fourier transform. Thus

$$(14) \qquad e^{-au} \Delta(e^u) = \frac{1}{2\pi i} \int_{-\infty}^{\infty} \frac{\zeta^2(a + it)}{a + it} e^{iut} dt.$$

Before proceeding further, we shall show that $\zeta^2 (a + it)(a + it)^{-1}$ belongs to $L^2(-\infty, \infty)$. Once this has been established we can use the Parseval-Plancherel theorem to obtain the following equation:

$$(15) \qquad \begin{aligned} \int_{-\infty}^{\infty} e^{-2au} \Delta^2(e^u)\, du &= \frac{1}{2\pi} \int_{-\infty}^{\infty} \frac{|\zeta(a + it)|^4}{a^2 + t^2}\, dt \\ &= \frac{1}{\pi} \int_{0}^{\infty} \frac{|\zeta(a + it)|^4}{a^2 + t^2}\, dt. \end{aligned}$$

To prove the L^2-result, we require the following three results.

$$(16) \qquad \int_{1}^{T} |\zeta(\sigma + it)|^4 dt \sim c_2(\sigma)T, \quad 1 > \sigma > \frac{1}{2} \quad (T \to \infty),$$

$$(17) \qquad \zeta(s) = 2^s \pi^{s-1} \sin \frac{\pi s}{2} \Gamma(1 - s)\zeta(1 - s).$$

$$(18) \qquad |\Gamma(1 - a - it)| \sim c_3(a)|t|^{\frac{1}{2}-a} e^{-\pi|t|/2} \quad \frac{1}{4} \leqslant a \leqslant \frac{1}{2} \quad (|t| \to \infty).$$

Combining (18) and (17), we obtain

(19) $|\zeta(a + it)| \sim c_4(a)|t|^{\frac{1}{2}-a}|\zeta(1 - a - it)|$ $(|t| \to \infty)$.

Thus the convergence of the right side of (15) is dependent upon the convergence of

(20) $$\int_0^\infty \frac{t^{4(\frac{1}{2}-a)}|\zeta(1 - a - it)|^4}{a^2 + t^2}\,dt,$$

which in turn is convergent if

(21) $$\int_1^\infty \frac{|\zeta(1 - a - it)|^4}{t^{4a}}\,dt < \infty.$$

We now integrate by parts, transforming (21) into

(22) $$\frac{\int_1^t |\zeta(1 - a - it)|^4\,dt}{t^{4a}}\Bigg]_1^\infty + 4a \int_1^\infty \frac{\left\{\int_1^t |\zeta(1 - a - it)|^4\,dt\right\}}{t^{1+4a}}\,dt.$$

Using (16) we see that the integrated part vanishes, and the integral is less than

(23) $$c_2 \int_1^\infty \frac{t\,dt}{t^{1+4a}} = \frac{c_3(a)}{a - 1/4},$$

provided $a > \frac{1}{4}$ and $< \frac{1}{2}$. This permits us to assert that (15) is valid.

We now retrace our steps in the preceding proof, and establish the asymptotic relation

(24) $$\int_0^\infty \frac{|\zeta(a + it)|^4}{t^2}\,dt \sim \frac{c_5}{a - 1/4}   \quad (a \to \frac{1}{4}).$$

Once this has been established, we shall have the following relation

(25) $$\int_{-\infty}^\infty e^{-2(a-1/4)u}e^{-u/2}\Delta^2(e^u)\,du \sim \frac{c_6}{a - 1/4}   \quad (a \to \frac{1}{4}).$$

It is easy to see that the integral over the interval $(0, -\infty)$ is finite as $a \to \frac{1}{4}$, and thus

(26) $$\int_0^\infty e^{-2(a-1/4)u}e^{-u/2}\Delta^2(e^u)\,du \sim c_6/(a - 1/4)$$

or

$$(27) \qquad \int_0^\infty e^{-su} e^{-u/2} \Delta^2(e^u)\,du \sim c_7/s \qquad (s \to 0).$$

The use of a familiar Tauberian theorem, due to Hardy and Littlewood, (27) leads to

$$(28) \qquad \int_0^T e^{-u/2} \Delta^2(e^u)\,du \sim c_7 T \qquad (T \to \infty),$$

whence a change of variable yields (7).

We now have to prove the relationship given in (24). It is again sufficient to consider the behavior of (21). It is now important to note that since (16) is valid for $\frac{1}{2} < \sigma < 1$, we have

$$(29) \qquad \int_1^T |\zeta(1 - a - it)|^4 dt = c_2\left(\frac{3}{4}\right)T + \epsilon T \qquad (\epsilon \text{ small}),$$

for a close to $\frac{1}{4}$ and T large. Here we have used the fact that $c_2(a)$ depends continuously upon a, which follows from the form of $c_2(a)$.

Using (29) and integrating by parts as in (22), we obtain the desired result (24) and the proof is complete.

We now consider the function defined in (9). The Ω-result we obtain here, and more, was obtained by Chowla and Pillai, using very different methods, but the mean-value theorem seems to be new.

Consider the generating function for the Euler ϕ-function,

$$(30) \qquad \frac{\zeta(s - 1)}{\zeta(s)} = \sum_{k=1}^{\infty} \frac{\phi(k)}{k^s} \qquad (R(s) > 2).$$

Applying the Perron sum-formula,

$$(31) \qquad \sum_{n<x} \phi(n) = \frac{1}{2\pi i} \int_{3-i\infty}^{3+i\infty} \frac{\zeta(s - 1)}{\zeta(s)} \frac{x^s}{s}\,ds.$$

Shifting the line of integration to the line $1 + a + it$, $0 < a < 1$, we obtain the equation

(32)
$$\sum_{n<x} \phi(n) - \frac{3x^2}{\pi^2} = \frac{1}{2\pi} \int_{-\infty}^{\infty} \frac{\zeta(a+it)}{\zeta(1+a+it)} \frac{x^{1+a+it}}{(a+it)} dt.$$

Denoting the left side of (32) by $\Delta_1(x)$, we obtain as before

(33)
$$\int_{-\infty}^{\infty} e^{-2(a+1)u} \Delta_1^2(e^u) du = \frac{1}{2\pi} \int_{-\infty}^{\infty} \left| \frac{\zeta(a+it)}{\zeta(a+1+it)} \right|^2 \frac{dt}{a^2+t^2} .$$

Using (19) and a bound on $|\zeta(1+a+it)|^{-1}$, we have

(34)
$$\left| \frac{\zeta(a+it)}{\zeta(1+a+it)} \right|^2 = O(t^{1-2a}(\log|t|^4)).$$

Thus the right side of (33) exists and, using Plancherel's theorem, (33) is valid.

To determine the asymptotic behavior of the left side of (33) as $a \to 0$, we have to consider the behavior of

(35)
$$\int_1^{\infty} \frac{|\zeta(1-a-it)|^2}{|\zeta(1+a+it)|^2} \frac{dt}{t^{2a+1}} .$$

We expect that as $a \to 0$, this acts like

(36)
$$\int_1^{\infty} \frac{dt}{t^{2a+1}} = \frac{c_8}{a} \quad (a \to 0).$$

To show this, we use the following equation.

(37)
$$\int_1^{\infty} \frac{|\zeta(1-a+it)|^2}{|\zeta(1+a+it)|^2} \frac{dt}{t^{2a+1}}$$

$$= \int_1^{\infty} \frac{dt}{t^{2a+1}} + \int_1^{\infty} \frac{(|\zeta(1-a+it)|^2 - |\zeta(1+a+it)|^2)}{|\zeta(1+a+it)|^2} = \frac{dt}{t^{2a+1}}$$

$$= \frac{1}{2a} + \int_{1-a}^{1+a}$$

$$\cdot \left[\int_1^{\infty} \frac{\{\zeta(1-a-it)\zeta'(\sigma+it) + \zeta(1+a+it)\zeta'(\sigma-it)\}}{t^{2a+1}|\zeta(1+a+it)|^2} dt \right] d\sigma.$$

Consider the integral; it is less than

(38)
$$2 \int_{1-a}^{1+a} \left[\int_{1}^{\infty} \frac{|\zeta(1-a-it)||\zeta'(\sigma+it)|}{t^{2a+1}|\zeta(1+a+it)|^2} dt \right] d\sigma,$$

which in turn is less than

(39)
$$\int_{1-a}^{1+a} \left[\int_{1}^{\infty} \frac{\zeta(1-a-it)^2}{\zeta(1+a+it)} \frac{dt}{t^{2a+1}} + \int_{1}^{\infty} \frac{\zeta'(\sigma+it)}{\zeta(1+a+it)}^2 \frac{dt}{t^{2a+1}} \right] d\sigma.$$

Applying Schwarz's inequality again, we obtain the following bound for the integral:

(40)
$$\int_{1-a}^{1+a} \left[\int_{1}^{\infty} \frac{|\zeta(1-a-it)|^4}{t^{2a+1}} dt + 2 \int_{1}^{\infty} \frac{dt}{|\zeta(1+a+it)|^4 t^{2a+1}} \right.$$
$$\left. + \int_{1}^{\infty} \frac{|\zeta'(\sigma+it)|^4}{t^{2a+1}} dt \right] d\sigma.$$

It is easy to show that the mean values

(41)
$$\lim_{T\to\infty} \frac{1}{T} \int_{1}^{T} |\zeta(1-a-it)|^4 dt,$$
$$\lim_{T\to\infty} \frac{1}{T} \int_{1}^{T} \frac{dt}{|\zeta(1+a+it)|^4},$$
$$\lim_{T\to\infty} \frac{1}{T} \int_{1}^{T} |\zeta'(\sigma+it)|^4 dt$$

exist, uniformly in a and σ for a positive and close to 0 and σ close to 1, and thus that the inner integrals of (40) are uniformly bounded by c/a as $a \to 0$, where c is a positive constant. Thus the entire expression of (40) is bounded by a constant as $a \to 0$. This proves the relation stated in (36).

Thus we obtain the following result:

Let

(42)
$$\Delta_2(x) = \sum_{n<x} \phi(n) - 3x^2/\pi^2.$$

Then

(43)
$$\int_{1}^{T} \frac{\Delta_2^2(x)}{x^3} dx \sim c \log T \quad (T \to \infty).$$

Hence, $\Delta_2(x) = \Omega(x)$ as $x \to \infty$.

Chowla and Pillai obtained the following results.

$$(44) \qquad \int_1^T \Delta_2(x)\,dx \sim cT^2 \qquad (T \to \infty)$$

and

$$(45) \qquad \Delta_2(x) = \Omega(x \log\log\log x) \qquad (x \to \infty).$$

In conclusion, we may remark that the method illustrated above furnishes a systematic means of obtaining both asymptotic results for the average of the error term and Ω-results.

6. The Mean Value of $|\zeta(\tfrac{1}{2} + it)|^4$

The procedure used in Sec. 4 may be used to obtain the mean value of $|\zeta(\tfrac{1}{2} + it)|^4$ naturally, the analysis is more formidable.

As we point out in the references, the mean value has been obtained by several authors using different methods.

For the method followed in Sec. 4, we require a functional equation due to Wigert. References to Wigert's original paper and a much shorter proof by Landau are given at the end of the chapter.

7. The Erdös-Selberg Method for the Prime Number Theorem

Let us say a few words about the famous result of Erdös-Selberg concerning the prime number theorem. Many mathematicians have believed that it was impossible to find a proof of the prime number theorem without leaving the real line.

The Erdös-Selberg proof consists of two steps. First, classical methods are used to derive a nonlinear recurrence relation. Then a Tauberian theorem is used to obtain from this relation the prime number theorem.

Miscellaneous Exercises

1. By using the relation $\int_R \zeta(s)\zeta(1-s)\,ds = 0$, where R is a suitably chosen rectangle, can we derive the asymptotic behavior of $\int_0^T |\zeta(\tfrac{1}{2} + it)|^2 dt$?

2. Consider the series

$$\sum_{n=1}^{\infty} \frac{(d(n) - \log n - (2c - 1))}{n^s}$$

so that this series converges in the critical strip and obtain an estimate for the line of convergence.

3. Show that this can be used to obtain an estimate for $\zeta(\tfrac{1}{2} + it)$.

4. Can the higher divisor functions be used in the same way?

Bibliography and Comments

Section 1

The general theory was done by Wiener. See

Wiener, N., *Tauberian Theorems,* Ann. of Math. 3̃3 (1) (1932), 1-100.

See also

Erdös, P., B. Gordon, L. A. Rubel, and E. G. Straus, *Tauberian Theorems for Sum Sets,* Acta Arith. 9 (1964), 177-189.
Erdös, P., and A. E. Ingham, *Arithmetical Tauberian Theorems,* Acta Arith. 9 (1964), 341-356.
Jukes, K. A., *Tauberian Theorems of Landau-Ingham Type,* J. London Math. Soc. 8 (2) (1972), 570-576.
Segal, S. L., *A Tauberian Theorem for Dirichlet Convolutions,* Illinois J. Math. 13 (1969), 316-320.

A shortproof of the Hardy-Littlewood tauberian theorem by Karamata is given in the book by Titchmarsh.

Titchmarsh, E. C., *Theory of Functions,* Oxford University Press, London, 1939.

Section 2

It is possible to find Tauberian theorems with an error term. This requires involved analysis, and, usually, a better result can be obtained from other types of analysis.

Section 3

See the books

Doetsch, G., *Guide to the Applications of the Laplace Transforms,* Van Nostrand-Reinhold, New York 1961.
——, *Introduction to the Theory and Application of the Laplace Transformation,* Springer-Verlag, New York, 1974.
Widder, D. V., *The Laplace Transform,* Princeton University Press, Princeton, New Jersey, 1955.

See

Buschman, R. G., *Asymptotic Expressions for $\Sigma_n^a f(n) \log^r n$,* Pacific J. Math. 9 (1959), 9-12.

Section 4

This method is due to Titchmarsh. See

Titchmarsh, E. C., *The Zeta-Function of Riemann*, Cambridge University Press, London, 1930.

    See

Putnam, C. R., *On Averages of the Riemann Zeta-Function*, Arch. Math. **11** (1960), 346–349.

In the book by Titchmarsh, cited in Chap. 5, other methods for finding the mean value of the square are given. In addition he gives references to various methods for finding the mean value of the fourth power. No simple method seems to exist for finding these mean values.

Titchmarsh's method can be used for the mean value of the fourth power with some additional effort. See;

Bellman, R., *Wigert's Approximate Functional Equation and the Riemann Zeta-Function*, Duke Math. J. **16** (1949), 547–552.

Section 5

This follows

Bellman, R., *The Dirichlet Divisor Problem*, Duke Math. J. **14** (1947), 411–417.

    See

Chowla, S. D., and S. S. Pillai, *On the Error Terms in Some Asymptotic Formulae in the Theory of Numbers* (1), J. London Math. Soc. **5** (1930), 95–101.
Hardy, G. H., *The Average Order of the Arithmetical Functions* $P(x)$ *and* $\Delta(x)$, Proc. London Math. Soc. **15** (2) (1916), 1–25.
——, *On Dirichlet's Divisor Problem*, Proc. London Math. Soc. **15** (2) (1916), 1–25.
Hardy, G. H. and J. E. Littlewood, *Contributions to the Theory of the Riemann Zeta-Function and the Theory of the Distribution of Primes*, Acta Math. **4** (1918), 119–196.
Ingham, A. E., *The Distribution of Prime Numbers*, Cambridge University Press, 1932.
Titchmarsh, E. C., "The Mean-Value of the Zeta-Function of the Critical Line," *Proc. of the London Mathematical Society (2)*, Vol. 27 (2) (1927/28), pp. 137–150.
——, *The Zeta-Function of Riemann*, Cambridge University Press, London, 1930.

Section 6

The details are given in

Bellman, R., *Wigert's Approximate Functional Equation and the Riemann Zeta-Function,* Duke Math. J. **16** (4) (1949), 547–552.

Other methods may be found in

Atkinson, F. V., *The Mean Value of the Zeta-Function on the Critical Line,* Proc. London Math. Soc. **47** (2) (1942), 174–200.
Hardy, G. H. and J. E. Littlewood, *On Lindelof's Hypothesis concerning the Riemann Zeta-Function,* Proc. Roy. Soc. London Ser. A **103** (1923), 403–412.
Ingham, A. E., *Mean-Value Theorems in the Theory of the Riemann Zeta-Function,* Proc. London Math. Soc. **27** (2) (1928), 273–300.
Titchmarsh, E. C., *The Zeta-Function of Riemann,* Cambridge Tract No. 26, Cambridge University Press, London, 1930.

Wigert's functional equation is given in his paper
Wigert, S., *Sur la série de Lambert et son application à la théorie des nombres,* Acta Math. **41** (1918), 197–218.

Section 7

A more direct method uses the Riccati equation.
Another procedure was given by Wright.
This method was extended to primes in an arithmetic progression by H. N. Shapiro.

Name Index

Name Index

Subject Index

Subject Index